AF879191

Printed in the United States of America
First Printing, 2023

ISBN: 978-1-959640-04-2

Modern Warrior Publications:
Division of Omega Group Consultants LLC.
490 County Road 39,
Headland, AL. 36345
modernwarriorproject.com

Table of Contents

Okinawa, The Bloody Perils of War

This book is dedicated to The Armed Forces of The United States of America who forever Hold the Torch of Liberty High.

Picture: The Infantry Wall, National Infantry Museum, Fort Benning, Georgia

Sergeant Rathel served in the 77Th Infantry Division in World War II. The Motto of the 77th Infantry Division is "Ours to hold it High," inspired by the third stanza in the poem, "In Flanders Fields," by John McCrae.

Shoulder Sleeve Insignia of the Legendary 77th Infantry Division

Statement from the Author

The one account left out of the many histories, commentaries, and documentaries, about The Battle of Okinawa, is the personal account of the many American Soldiers who participated in the battle. Their personal experiences in the battle and their expressions of reconciliation with the people of Japan and the people of Okinawa are quickly fading into history untold. I can honestly say that my Father, Sergeant First Class Clarence Rathel who served on the front lines in the Battle of Okinawa never said one unkind word about the people of Japan or the people of Okinawa. I know his heart and he was always compassionate to both people. As a Soldier representing our nation, his greatest ideal was Liberty for the oppressed people of the world and his place in the fight for Liberty should be told. As time passes and history vanishes there are many lessons we should all heed from those who experienced the Battle of Okinawa, possibly the bloodiest battle in World History. I feel honored to share ***his story.*** *This personal account also includes many of the historical events that occurred during the time of my Father's service in World War II, including the occupation of Japan after the war.*

All Italicized words are from documented resources as noted or written by the author for background history and clarity.

Appreciation and Acknowledgements

I would like to extend my heartfelt appreciation to those who have encouraged me on this adventure of writing and research, especially my wife Denise, without her unending encouragement and love, this series of books would not be possible. I have also been blessed with motivation from my publisher Shane Blevins and his amazing publishing company Modern Warrior Publications. Shane has served our Nation on the frontlines and his understanding of the military spirit is a virtuous gift.

One of the greatest blessings on this journey has been to meet those who volunteer at the many museums and memorial parks we have visited; their Patriotism and knowledge are like a spark that ignites history and brings it to life. Their love for our country is seen in their kindness, smiles, and service.

I would like to give a special thank you to the Woodring Wall of Honor and Veterans Park, Inc. Enid, Oklahoma, for allowing me to share many of the pictures recently discovered in an old suitcase containing notes and letters of a deceased Army Veteran who served in The Battle of Okinawa; this find was truly a treasure trove of information. The discovery of this historical treasure beckons me to remind everyone to please donate their loved ones items of patriotic memories to museums.

I would also like to thank and commend The National Infantry Museum, Fort Benning, Georgia, for its phenomenal Tribute to our Nation's Army and American History. I was struck again and again with inspiration as I beheld the many displays that represented the battles my Father lived and encountered in the Pacific Campaign of World War II, and the Korean War. Growing up as a child in Columbus, Georgia our family lived the Fort Benning Army Infantry life experience.

I will forever be thankful for the majestic displays of military aircraft, aviation history, and tributes to our nation's aviators at the Museum of Aviation, Robins Air Force Base, Georgia. One display of our Nation's heroic aviators flying unarmed over the most dangerous terrain, the Burma Hump, possibly the greatest aviation challenge in the history of flight, really got my attention. From the historic B-29 Superfortress display to the display of our amazing Air Force Thunderbirds F-16 "Fighting Falcon," I was astonished.

I would like to thank the National Prisoner of War Museum, Andersonville, Georgia for the respect and honor given to our nation's Prisoners of War, their displays are truly heartfelt, and while visiting I was humbled with Patriotism. Its cemetery is immaculate, and its memorials touch your soul.

Upon arrival at Fort Mitchell National Cemetery, Fort Mitchell, Alabama I stood in awe as I gazed upon the memorials Honoring our fallen military, Gold Star Families, and The Memorial, "Remember The Rohna."

The National World War II Museum in New Orleans, Louisiana is a National treasure and has detailed displays, a Four D Movie, and an interactive submarine experience on the Navy's "USS Tang Submarine," sunk by her own torpedo in World War II. I recommend this museum as an unforgettable family experience.

The Columbus Museum, in Columbus, Georgia has in-depth historical displays that include Military, National, Civil Rights, and Southern History. We were in awe as we carefully studied its display,"

Steamboats on the Chattahoochee." Growing up in Columbus made me proud to see its history presented in a way that informs all ages.

The USS Alabama Battleship Memorial Park, Mobile, Alabama is a Naval and Military Treasure. The displays are perfectly restored and the USS Alabama's armament of nine 16 inch guns is breathtaking.

The on-site displays of military weapons, tanks and even the USS Drum Submarine are worth the visit. The parks memorials are soul touching and honor our fallen war heroes.

Thank you,
Randy Rathel

Introduction

My Father, Sergeant First Class Clarence Rathel considered his Military Oath of Enlistment a lifetime commitment. After he retired from the Army in 1964, at Fort Benning, Georgia, our family moved to Headland, Alabama; a town our family loved. As time passed I realized that a special bond existed between my Father and many of the leaders in our community. I was puzzled about this unique bond but never questioned it.

In book one I explained that several months before my Father passed away I asked him to record the story of his life; especially his military battles in World War II and the Korean War. It was my hope the process of recording would shift his mind away from the constant pain he endured from cancer. After his death, I placed the cassette recordings in my desk drawer and only recently revisited them. Enthralled by what he said, I began a serious study of his battles in World War II and the Korean War. It was during this study that I finally discovered the key to the bond my Father shared with the men of our town. It was a bond built upon mutual respect, for they too had humbly served our Nation in war and held "The Torch of Liberty High."

Before we continue please allow me to recap my Father's journey detailed in book one to where we begin the Battle of Okinawa in World War II.

In book one I shared the story of how my Father was raised and lived through the Great Depression of the 1930s on his family's farm in Gordon, Alabama. As the United States entered World War II at the age of twenty-four he was drafted; on November 4th, 1942, he became a soldier in the 77th Infantry Division assigned to Company A of the 307th Regimental Combat Team.

After intense training in the deserts of Arizona and California preparing for battle in North Africa there was a sudden mission change; his training shifted to the campaign in the Pacific Theater of War. After weapons, survival, mountain climbing, and amphibious assault training in the northeast United States, his division was sent to Hawaii for jungle warfare and martial arts training.

From July to September 1944 he participated in multiple battles in the Liberation of Guam. From October 1944 to February 1945, he participated in several fiery battles to Liberate the island of Leyte. After Leyte, his Division took part in the ***first battle*** *of Okinawa on March 26th, 1945, in the Kerama Islands. In the Kerama Islands, the 307th RCT faced enemy Banzai attacks and ultimately destroyed over 300 Imperial Japanese suicide boats. The Kerama Islands also provided a naval position for the main invasion of Okinawa on April 1st, 1945.*

After leaving the Kerama Islands his transport flotilla in the Pacific Ocean was attacked by a wave of kamikaze pilots. On the 16th of April while en route to the main island of Okinawa the 307th RCT was redirected to help the other combat teams of the 77th Infantry on the island of Ie Shima; arriving and making an amphibious assault on the 17th of April. The enemy was so heavily fortified on Ie Shima and the battle so fierce his regiment suffered many casualties. The battle of Ie Shima ended on the 25th of April, 1945.

Four days after the horrendous Battle of Ie Shima the 77th Infantry Division suffering from battle fatigue and undermanned entered the Battle of Okinawa on the 29th of April 1945. Their first assignment Hacksaw Ridge.

Chapter One

Okinawa, A Peal of Thunder

"The Battle of Okinawa," was one of the most important battles in history, its effect on Western Civilization and World History cannot be overstated. The outcome of how this battle ended ultimately led President Harry Truman to make one of the most dramatic decisions in history--The use of Atomic Weapons, on Japan.

Okinawa Campaign, Pre-Landing Bombardment, Last week of March to April 1, 1945
Picture: ***https://www.history.navy.mil/content/history/museums/nmusn/search.html?q=Lc-usz62-92435***

Pictures: The National World War II Museum, New Orleans, Louisiana

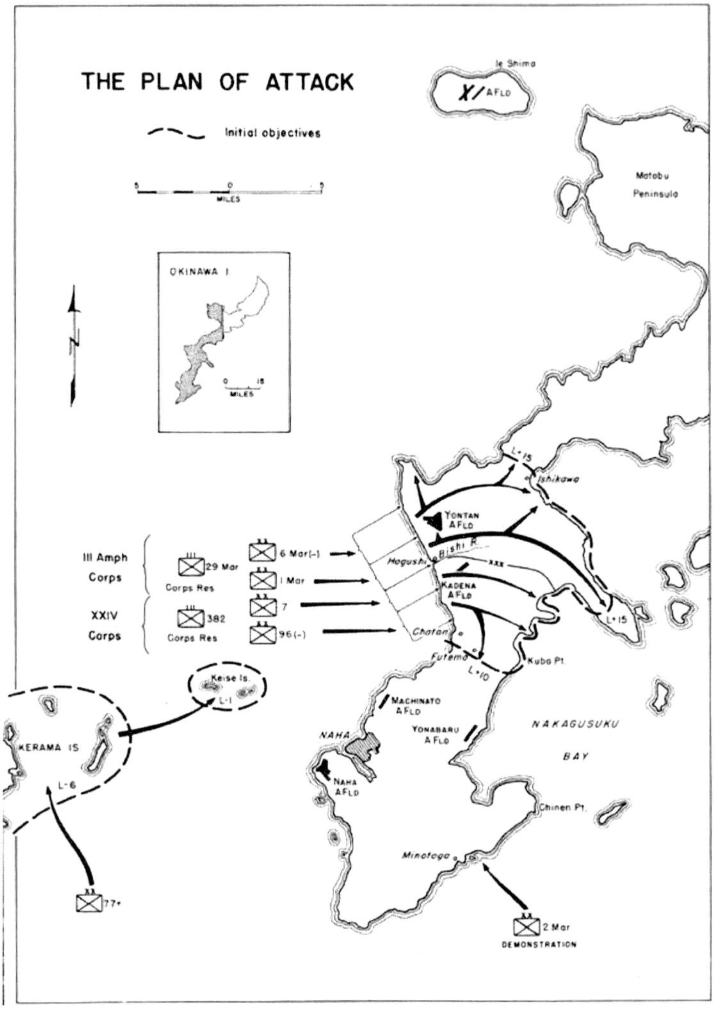

Map: history.army.mil: The U.S. Army in WW II, The War in the Pacific, Okinawa The Last Battle Map 3, The plan of Attack page 30. (The Tenth Army is represented by the diagonal box with an X.)

"A Date Which Will Live in Infamy,"

*As we consider the "Battle of Okinawa " we should remember the **promise** President Franklin D. Roosevelt made to The United States Congress and the American people in his Declaration of War speech the day after the Imperil Japanese attacked Pearl Harbor.*

"<u>No matter how long it may take</u> us to overcome this premeditated invasion the American people in their righteous might will win through to absolute Victory," December 8th, 1941, President Franklin D. Roosevelt.

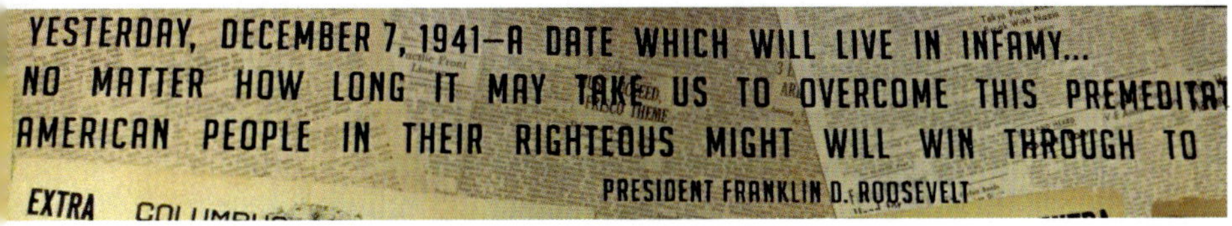

Picture from a newspaper display, The Columbus Museum, Columbus, Georgia

*The United States attack on the Kerama Islands on March 26, 1945, was the **first official** land invasion on Japanese soil during World War II, this happened **just over three years after the Pearl Harbor Attack. The 77th Infantry Division conducted the Kerama Islands invasion. Sergeant Rathel's 307th Regimental Combat Team was part of the invasion force.***

***The Battle of Okinawa** was codenamed, **"Operation Iceberg."** The battle was from **April 1 to June 22,** 1945. The Allies planned to defeat the Imperial Japanese on Okinawa and then use Okinawa as a base to attack Japan in a battle codenamed, "**Operation Downfall**." There will be more information about Operation Downfall, including a map as this book progresses.*

The Naval and Air Battles during the Battle of Okinawa are beyond our comprehension.

*The total Allied Forces involved were approximately 550,000,(*1) this number includes the Tenth Army's force of 183,000(*2) Soldiers and Marines. The United States Navy, Marines, and Army Air Force utilized a diverse fleet of aircraft, the total number is estimated at over 3,000. The number of Allied naval vessels is guessed to be near 2,000 and includes submarines. The United States Navy, Coast Guard, and Merchant Marines were essential for victory in this battle.*

*In the battle at sea, the Japanese sacrificed their largest battleship the Yamato in a suicide attack against the Allied fleet. Official military records show there were 1,465 Japanese Kamikaze suicide pilot attacks against the Allied fleet during the Battle of Okinawa. (*3) The Japanese submarine fleet during World War II was far more diverse than those of the Allies. The Imperial Japanese also used manned torpedoes and suicide boats, in the Battle of Okinawa.*

**1 The World War II Museum, New Orleans, Louisiana*
**2https://history.army.mil/brochures/ryukyus/ryukyus.htm Page 8, paragraph 1*
**3 https://history.army.mil/books/wwii/okinawa/chapter14.htm Page 364* (https://history.army.mil/books/wwii/okinawa/chapter14.htm Page 364), *paragraph 1*

Okinawa the Island

The distance from Okinawa to Japan's southernmost island of Kyushu is 400 miles. Okinawa is the largest island in the Ryukyu Islands, a group of islands also known as the Nansei-shoto. Okinawa is approximately sixty miles long and sixteen miles wide at its widest point. Made of coral and limestone the island has numerous caves on its coast and inland. The southern region of the island has a very rugged terrain with steep ridges. Its climate is subtropical and during the spring its daytime temperature averages between 70 and 80 degrees. The Battle of Okinawa was in spring. Okinawa has a monsoon season that occurs in May and June, the island is also subject to typhoons in the summer and fall.

The People of Okinawa

The indigenous people of Okinawa have an ancient culture and are called the Ryukyuan People as are the people of the other Ryukyuan islands. Okinawa was annexed by Japan in 1879; Japanese influence upon the government, history, and culture of Okinawa is crucial to understanding how The Battle of Okinawa affected its citizens before, during, and after the war. As we consider the final battle of World War II it should be remembered that it was one of the bloodiest conflicts in human history. The Battle of Okinawa brought not only human tragedy on a scale never before seen but the fragile island itself was devastated and changed forever. As the battle ended Okinawa's cities and villages were left in Apocalyptic ruin.

*Special note from the author: It would be impossible for me to cover in this writing the atrocities the people of Okinawa suffered before and during World War II. The truth is atrocities were committed by both the Japanese and Allied forces. In the end, it should be remembered the people of Okinawa were caught in the Crossfire with no escape, and **their beautiful island fractured by war.***

Fractured landscape above and the destroyed Music Hall in Naha, Okinawa, below: These two pictures were recently discovered in a donation containing the memories of a deceased Army veteran who had served in the Battle of Okinawa. The Woodring Wall of Honor and Veterans Park, Inc. Enid, Oklahoma.

The Victory Flag, the first American Flag to fly over occupied Japan after World War II
Picture of the actual flag: The National World War II Museum, New Orleans, Louisiana

Military Organization of The United States Tenth Army

*The 77th Infantry Division was part of the **XXIV Corps.** The 7th, 27th, 77th, and 96th, Army Infantry Divisions formed the **XXIV Corps. The XXIV Corps and the III Amphibious Marine Corps of the U.S. Marine Corps combined to form the Tenth Army.** The Tenth Army had its own Tactical Air Force and the Navy's Fifth Fleet. The Tenth army had approximately 183,000 troops that took part in various assault phases of the land battle.*

***The Tenth Army** invaded Okinawa on April 1st, 1945, (Easter Sunday) with more than 60, 000 (*4) soldiers and U.S. Marines in the largest Amphibious Assault in the Pacific during World War II. In late March, American B-29 bombers attacked the Japanese airfields in Kyushu, Japan. This attack minimized Japanese air attacks on the Tenth Army's amphibious assault landings on April 1. For seven days before the landings, the Navy bombarded key positions along Okinawa's coastline. On the morning of April 1st, before the beach landings, beginning at dawn over 1,600 (*5) Navy vessels began the heaviest pre-landing bombardment in support of an amphibious assault in the Pacific Theater of War during World War II. Once on the beaches, the Tenth Army experienced minimal enemy opposition. The enemy's strategy was to bring the battle inland almost to the center of the island where their defenses were extremely fortified; this move also allowed the enemy to avoid naval bombardment on the coast; a ploy they had learned from the battle of Saipan.*
*(*4)history.army.mil: Title Ryukyus, page 10, paragraphs 4, and 5*
*(*5)Command And Control Of The U.S. Tenth Army During The Battle of Okinawa: apps:dtic.mil/sti/pdf/ADA501755.pdf page14 paragraph*

The War Flag of the Imperial Japanese Army, The Flag of the Rising Sun

The Imperial Japanese, Thirty-Second Army

*The Imperial Japanese Thirty-second Army included their Naval Fleet and Air support. The estimated manpower for the Imperial Japanese shifts between 115,000 and 116,000,(*6)* ***in the Battle of Okinawa, true figures are a guess because the Imperial Japanese conscripted the citizens of Okinawa, including school children. It's important to note that a portion of the population of Okinawa identified themselves as Japanese and supported the Imperial Japanese in the war.***
*(*6)Estimated numbers from different historical articles*

The advantage the Imperial Japanese had in Okinawa was their unique fortifications on ***escarpments,(*7)*** *ridges, and hills; their use of fortified honeycombed tunnels allowed weapons to be hidden in caves having multiple secret levels with well-concealed openings...A year before the battle the Japanese started preparing to defend the island. They created minefields, tank traps, and fortified machine-gun pill boxes; they even converted burial tombs into machine-gun pill boxes. They also used* ***self-sacrifice as offensive and defensive weapons. Unfortunately, surrender was not considered an option for most of the Imperial Japanese.***

The following is a statement from the National World War II Museum in New Orleans, Louisiana:

"Japan's warriors were indoctrinated from youth to embrace a militaristic code that exalted dying for the emperor and Japanese racial superiority, while emphasizing the shamefulness of surrender. In China and elsewhere, Japanese forces often subjected non-Japanese peoples, soldiers and civilians alike, to barbaric treatment. Their use of mass suicide attacks, known as banzai charges, seemed to divorce the Japanese from a shared humanity. To the shocked Allies, the Japanese seemed to demand a war without mercy, and Allied soldiers often answered in kind."

*(*7)Escarpment- A steep slope in front of a fortification. Merriam-Webster Dictionary*

Below is an excerpt of a document showing Sergeant Rathel's service in Company "A" 307TH Infantry. He served in the Southern Philippines, Western Pacific, and "RYUKYUS" (the Kerama Islands and Okinawa are part of the RYUKYUS ISLANDS.)

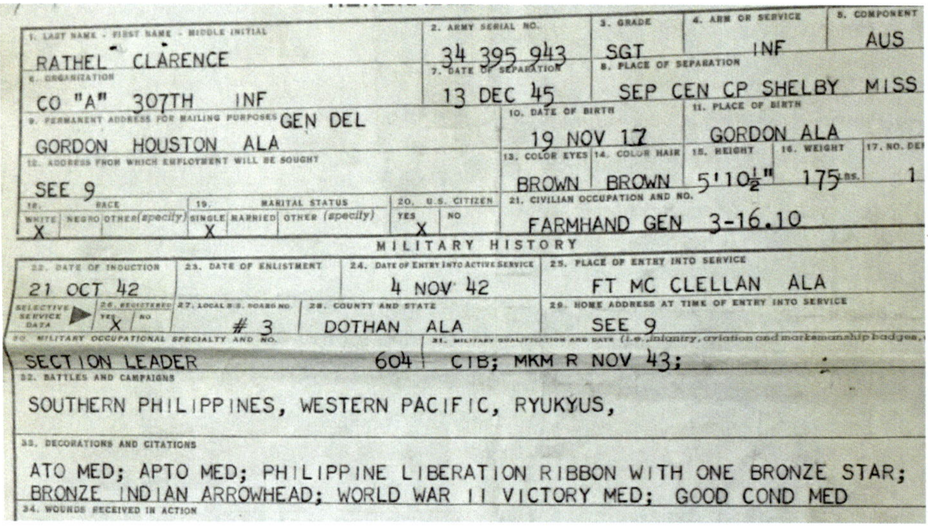

1. LAST NAME - FIRST NAME - MIDDLE INITIAL: RATHEL CLARENCE
2. ARMY SERIAL NO.: 34 395 943
3. GRADE: SGT
4. ARM OR SERVICE: INF
5. COMPONENT: AUS
6. ORGANIZATION: CO "A" 307TH INF
7. DATE OF SEPARATION: 13 DEC 45
8. PLACE OF SEPARATION: SEP CEN CP SHELBY MISS
9. PERMANENT ADDRESS FOR MAILING PURPOSES: GEN DEL GORDON HOUSTON ALA
10. DATE OF BIRTH: 19 NOV 17
11. PLACE OF BIRTH: GORDON ALA
12. ADDRESS FROM WHICH EMPLOYMENT WILL BE SOUGHT: SEE 9
13. COLOR EYES: BROWN
14. COLOR HAIR: BROWN
15. HEIGHT: 5'10½"
16. WEIGHT: 175 LBS.
17. NO. DEP.: 1
18. RACE: WHITE X
19. MARITAL STATUS: SINGLE X
20. U.S. CITIZEN: YES X
21. CIVILIAN OCCUPATION AND NO.: FARMHAND GEN 3-16.10

MILITARY HISTORY

22. DATE OF INDUCTION: 21 OCT 42
23. DATE OF ENLISTMENT:
24. DATE OF ENTRY INTO ACTIVE SERVICE: 4 NOV 42
25. PLACE OF ENTRY INTO SERVICE: FT MC CLELLAN ALA
SELECTIVE SERVICE DATA — 26. REGISTERED: YES X
27. LOCAL S.S. BOARD NO.: # 3
28. COUNTY AND STATE: DOTHAN ALA
29. HOME ADDRESS AT TIME OF ENTRY INTO SERVICE: SEE 9
30. MILITARY OCCUPATIONAL SPECIALTY AND NO.: SECTION LEADER 604
31. MILITARY QUALIFICATION AND DATE (i.e., infantry, aviation and marksmanship badges, etc.): CIB; MKM R NOV 43;
32. BATTLES AND CAMPAIGNS: SOUTHERN PHILIPPINES, WESTERN PACIFIC, RYUKYUS,
33. DECORATIONS AND CITATIONS: ATO MED; APTO MED; PHILIPPINE LIBERATION RIBBON WITH ONE BRONZE STAR; BRONZE INDIAN ARROWHEAD; WORLD WAR II VICTORY MED; GOOD COND MED
34. WOUNDS RECEIVED IN ACTION

In Okinawa the 77th Infantry Division's 307th Regimental Combat Team, (RCT) consisted of three Battalions plus artillery support and medical, or approximately 3,100 men. The 307th RCT first battled on the Madea Escarpment (Hacksaw Ridge) then Chocolate Drop Hill, and then the Three Sisters (three hills) near Shuri Castle. The 307th was in the center with the 305th RCT and the 306th RCT on each side. These Regimental Team's rotated positions at times for rest. The 77th Infantry Division was part of The XXIV Corps which was part of the Tenth Army.

Map showing major battles in The Battle of Okinawa, please notice the location of Shuri Castle, Imperial Japanese Headquarters, also please notice the ship and airplane icons showing kamikaze attacks: The National World War II Museum, New Orleans, Louisiana

Chapter Two

When Bullets Fly - The Madea Escarpment

Madea Escarpment viewed from the east end of Kazazu Ridge on Okinawa
Picture: History.Army.Mil: Chapter XI: Assaulting the Second Suri Defense Ring: Page 277

Near the eastern end of the Urasoe-Mura Escarpment is a 350-foot sheer cliff called the "Madea Escarpment" named after the village of Madea which lies on its south slope; it is also called "Hacksaw Ridge" and the "Big Escarpment." The hill mass centering on the eastern end of the Urasoe-Mura Escarpment was often called Hill 196 in the official reports.(*8)
*(*8)History.Army.Military- Chapter XI: Assaulting the Second Suri Defense, Paragraph 2, Page 274*

<u>(The area to the right of Needle Rock is The Big Escarpment - Hacksaw Ridge, the machine-gun pillbox was positioned in a tunnel opening.)</u>

The Shuri Line and Shuri Castle

Before the arrival of the 77th Infantry Division on the 29th of April, the Imperial Japanese had concentrated their forces on the southern end of Okinawa and formed a line of powerful defense called the Shuri Line. Behind the Shuri Line was a heavily-fortified zone that stretched from coast to coast across very rugged terrain, connecting the towns of Yonabaru, Shuri, and Naha the capital city. For the Japanese, this zone formed an interlocking triangle of defense. In the center of this defensive triangle was Shuri Castle, a fortress having underground caverns with a network of tunnels. Shuri Castle was used by the Japanese as their headquarters and became a major focal point in the Battle of Okinawa. Shuri Castle is located three miles north of Naha and is protected by fortifications on Chocolate Drop Hill, Ishimmi Ridge, and Flat Top Hill all less than a mile and a half away. These landscape boundaries formed an impregnable defensive triangle in front of Shuri Castle. Before our

Army could approach Shuri Castle from the north, it had to overcome ***The Madea Escarpment known as Hacksaw Ridge.***

Sergeant Rathel speaking,

April 29th – May 7, 1945

After the battles on the Kerama Islands and Ie Shima, the 77th Infantry Division arrived in Okinawa on the 29th of April, 1945; our mission was to replace the 96th Infantry Division, which had been in battle since the invasion on April 1st. Upon our arrival, the 307th RCT replaced the 381st Infantry Regiment at the Madea Escarpment. Before our arrival, the 381st had experienced an extremely high number of wounded and killed in a fierce battle that had raged day and night for several days.

(From the official report)
"By the time it was relieved, the 381st had been reduced to about 40 percent combat efficiency and had suffered 1,021 casualties, 536 of them in the Maeda Escarpment fighting of the past four days. Some platoons were down to five or six men."(*9)
*(*9) history.army.mil chapter Xi: Assaulting The Second Shuri Defense Ring paragraph 4, Page 279*

The area in front of the escarpment offered no ground cover for our soldiers. From the ridge top, the enemy observed our every move, relaying it to their Headquarters at Shuri Castle. No matter what we tried on the ground there was a counterattack. On the opposite slope of the ridge, the enemy had built a fortress in caves connected with honeycomb tunnels throughout the escarpment; providing cover and concealment for their weapons. The enemy also had mortar and machine gun pillboxes protected by concrete barriers. At night enemy snipers attacked from camouflaged openings in the limestone ridge with a sudden burst of machine gun fire. Sometimes snipers attacked from our rear in stealth from tunnels in the ground.

In this battle, The Tactical Air Force showered the escarpment daily with waves of napalm, bombs, and machine gun fire. All of our big guns were used from tanks with flame throwers to 155 mm Howitzers. Often there were hand grenade battles with us throwing hand grenades as fast as we could, of course, the enemy did the same. In this battle our soldiers cried from battle fatigue, lacking food, water, and sleep.

My Father said this was one of the hardest-fought battles he remembered, in fact, he was emotional as he spoke of this battle. For that reason, I will share with you the final paragraphs about Hacksaw Ridge from Army History.

(The following is from the official report)

*"The fighting was **especially fierce on top of the escarpment** on 4 May, Colonel Cooney's 1st Battalion successfully executed a complicated demolition assault on the big cave-tunnel-pillbox network about 200 feet west of the eastern end of the escarpment. The battalion then held the newly won ground against repeated counterattacks from the southern side. It was estimated that 600 enemies were killed by the 307th Infantry, in the fighting at the escarpment on 4 May. Slowly, on 5 May, the reverse slope was taken and caves blasted and closed. On the night of 5-6 May the Japanese staged several counterattacks to win back the escarpment. An especially severe attack struck the 3rd Battalion, 307th Infantry, on the Regimental right. In repelling it the 3rd Battalion killed 250 enemy troops, **largely in hand-to-hand** fighting. On 6 May all battalions of the 307th Infantry advanced southward to the slopes of Hill 187. The battle of the escarpment was over."(*10)*
*(*10) History.Army.Mil Chapter XI: Assaulting the Second Shuri Defense Ring: Maeda Escarpment Falls, Paragraph 2, Page 281*

Important
"The losses in the escarpment battle had been heavy. In the 1st Battalion, 307th Infantry, no less than eight company commanders were wounded in one 36-hour period. It had gone up on the escarpment on 29 April with a strength of about 800 men; it came down on 7 May with 324. The 77th Division estimated that it had killed upwards of 3,000 Japanese in the 7-day battle for the escarpment.."(*11)
*(*11) History.Army.Mil Chapter XI: Assaulting the Second Shuri Defense Ring: Maeda Escarpment Falls, Page 281 Paragraph 4, Page 282 Paragraph 1*

******The escarpment battle ended on May 7th, on May 8th, 1945 there was Victory in Europe.******

American Army Soldier Imperial

Japanese Sniper in Uniform

Pictures: National Infantry Museum, Fort Benning, Georgia

Chapter Three

The Fall of the Third Reich

May 8th, 1945

Picture: The World War II Globe, The National Infantry Museum, Fort Benning, Georgia

******As the escarpment battle -Hacksaw Ridge -ended on May 7th, on May 8th, 1945 there was Victory in Europe.******

The following statement is from The National World War II Museum, New Orleans, Louisiana.

"Allied troops advancing into Germany were surrounded by unimaginable violence and destruction. The drive to eliminate pockets of enemy resistance brought devastation to cities, towns, and civilians already battered by countless Allied air raids. With Germany's industry crippled and it's economy collapsing. Hitler ordered his followers to fight to the death. But the German army began to collapse, it's decimated ranks increasingly filled with old men and boys. Allied forces pushed deep into the heart of Germany and Soviet troops encircled Berlin. Hitler committed suicide. With much of the country in ruins, Germany formally surrendered on May 8, 1945."

Chapter Four

Victory in Europe, May 8, 1945

Victory in Europe--President Truman calls for A Day of Prayer on Mother's Day May 13, 1945

Newspaper Headlines, The Columbus Museum, Columbus, Georgia

President Harry Truman dedicated Victory in Europe Day, May 8th to President Franklin Delano Roosevelt who had died only a month earlier on April 12, 1945; the Flags of the United States remained at half-staff for the remainder of the thirty-day mourning period. As President Truman addressed the Nation that day he emphasized that the war with Japan continued. President Truman also called for Mother's Day, May 13, 1945, to be a ***Day of Prayer****.*

Chapter Five

The Holocaust Revealed

There were more than 28,000 NAZI Concentration Camps at the end of World War II.

Annelies Marie "Anne" Frank, a Jewish girl kept a diary of her family's Jewish persecution in Amsterdam during the Nazi German occupation of the Netherlands in World War II. She and her sister Margot died in a NAZI concentration camp in March 1945.
Statue of Anne Frank- The National World War II Museum, New Orleans, Louisiana

"Never Forget" The Holocaust

*Soon after the Victory in Europe was announced the world was shocked as it heard of the horrific horrors of **The Holocaust**. Words like Death Camps, Concentration Camps, mass murder, mass burial sites, and genocide were used trying to describe how Adolf Hitler's Nazi German regime eliminated an estimated six million Jews and others they considered undesirable. Some of the largest extermination camps were Dachau, Buchenwald, Treblinka, and Auschwitz. **In sadness, the world asks the question, how could this happen?***

The caption: Victims were stripped of all personal belongings before entering a concentration or extermination camp. Heads were shaved so that human hair could be used; after death, gold was removed from teeth. These shoes and clothing were taken from victims of the Auschwitz extermination camp.

Picture: The National Infantry Museum, Fort Benning, Georgia

Pictures and statement: The National Infantry Museum, Fort Benning, Georgia

"Concentration Camp Prisoner's Jacket: Pajama style uniform and wooden shoes were issued to each individual, underwear and socks were not. Worn year-round, this clothing did little to protect the individual during the winter. Starting in 1938 Jewish prisoners were forced to wear a yellow star which was a perversion of the Star of David."

Unbelievable-More than 28,000 NAZI German Concentration Camps, by the end of the war! The map below shows only two percent.

Map and statement: The National Infantry Museum, Fort Benning, Georgia

The caption: The United States Army and other Allied forces were the first to encounter the gruesome sights and stench of the NAZI concentration camps. The map to your right illustrates about two percent of the camps established by the NAZIs which numbered more than 28,000 by the end of the war. Shown are the flags representing the United States Army Infantry and Armored Divisions that liberated many of the camps.

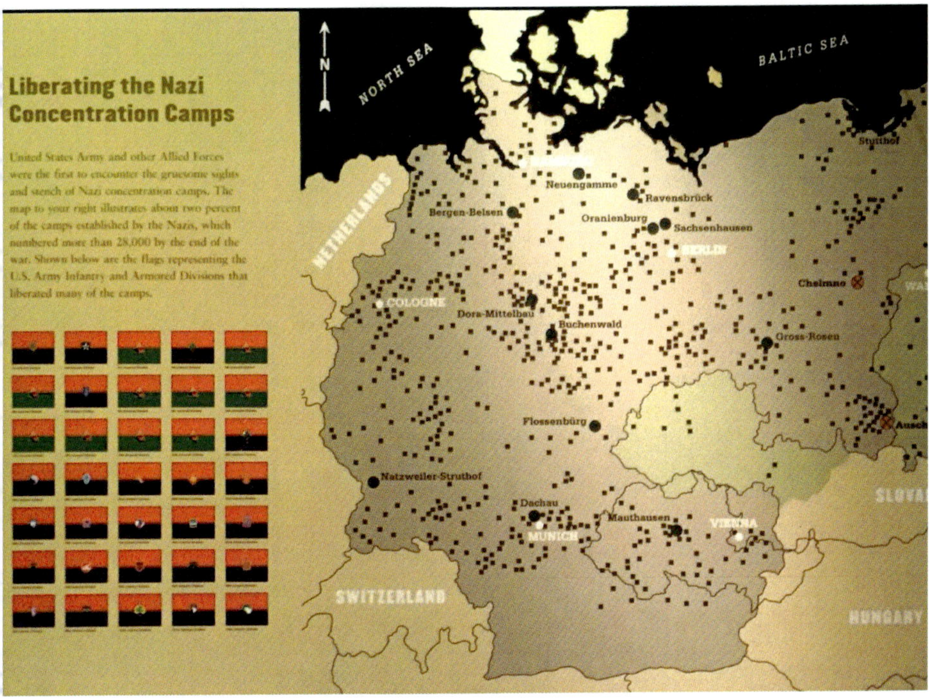

Chapter Six

The Liberation of NAZI Germany's Prisoners of War

*Also Liberated during the European Victory were NAZI Prisoners of War. Germany operated approximately one thousand Prisoner of War Camps during World War II, and imprisoned an estimated 94,000 Americans during the period from 1939 to the end of the war in 1945.(*13)*

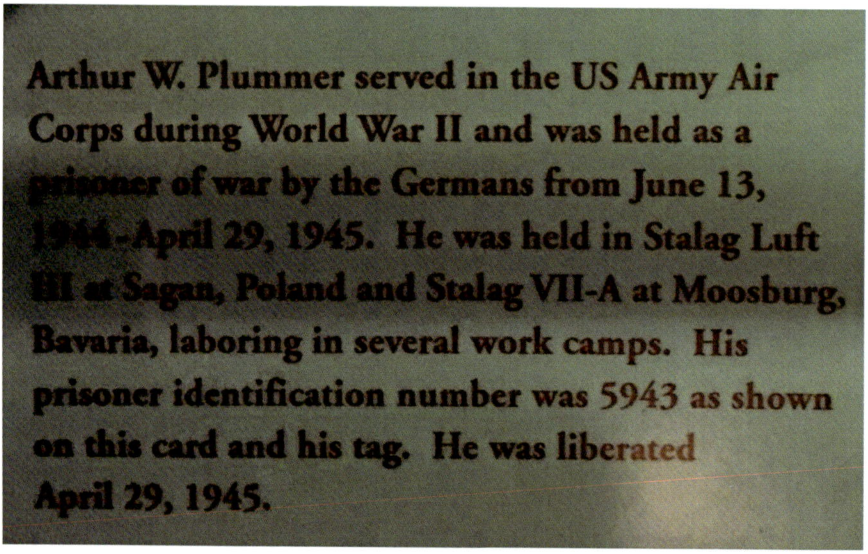

Arthur W. Plummer served in the US Army Air Corps during World War II and was held as a prisoner of war by the Germans from June 13, 1944-April 29, 1945. He was held in Stalag Luft III at Sagan, Poland and Stalag VII-A at Moosburg, Bavaria, laboring in several work camps. His prisoner identification number was 5943 as shown on this card and his tag. He was liberated April 29, 1945.

Picture: "The Georgia Monument," serves as a memorial to all American Prisoners of War, Andersonville National Cemetery, Andersonville, Georgia.

Picture and Statement: The National Prisoner of War Museum, Andersonville, Georgia
*(*13) navy.mil/research/library/online-reading-room/title-list-alphabetically/u/us-prisoners-war-civilian-american-citizens-captured.html*

Chapter Seven

Advance Toward Shuri

Imperial Japanese Soldiers Ready for Battle

Picture: The Nation World War II Museum, New Orleans, Louisiana

ɅIay 8th, 1945

ergeant Rathel, speaking

y May 8th the Tenth Army's plan of attack was to envelop the Shuri defenses. This would be ac-omplished by the Tenth Army's two corps, the Marine Divisions on the west and the Army Divi-ions on the east and center. The front line of attack was across the complete island advancing outhward into the heart of Japanese-controlled territory. The approach to Shuri was through very eacherous terrain. Shuri was on high ground surrounded by hills. These hills were used as a uffer protecting the Japanese Headquarters in Shuri. Many of the hills and ridges were fortifica-ons containing caves connected by tunnels used to supply weapons, ammunition, and fresh en-my soldiers. From the enemy's elevated position they had the advantage of observation. Knowing ur weak points they were able to attack in waves with heavy artillery, machine gun crossfire, and nortar fire. Our Division advanced slowly during the daylight hours but during the night the enemy sually tried to infiltrate our positions, sometimes they would send a soldier forward, provoking us o fire. Our response gave away our position then an intense battle would begin often leading to and-to-hand combat. There was no rest! It was very difficult just to eat our C-Rations and keep ur weapons clean. Above us daily there was a war in the heavens, Japanese Zeros, and our TAF Tactical Air Force) in dog fights. From our elevation, we also observed kamikaze attacks against our Navy.

Above: The escort carrier USS St. Louis (CVE-63) was the first major warship sunk by a kamikaze, struck by a bomb laden (A6M5) Zero fighter off Samar on October 25, 1944:

"Kamikaze – suicide plane attacks began during preliminary operations at Okinawa on March 26, 1945. Five days after the initial landings on April 1st, a wave of 355 Japanese army and navy kami-kaze aircraft struck the armada of Allied ships supporting the invasion, and further attacks contin-ued into June. By war's end, Japan would launch almost 2,000 suicide attacks against the Allied fleet, including manned rocket-powered Ohka flying bombs. The attacks tested the nerves of even veteran sailors as 26 ships were sunk and another 164 damaged."

Picture and statement: The National World War II Museum, New Orleans, Louisiana

Kamikaze – "Devine Wind Special Attack Unit", these Japanese pilots flew suicide attacks for Imperial Japan against the Allies.

Picture: The USS Alabama Battleship Memorial Park, Mobile, Alabama

May 9th, 1945

On the ridges approaching Shuri, our mission was to pinpoint and knockout enemy strongholds with heavy artillery attacks. Our main objective was to destroy and seal cave openings, burying the enemy with their weapons of death. The Japanese knew our intentions and after one fiery battle, we sealed a cave opening with a tank flamethrower. Suddenly deep within the cave explosions were heard as the enemy killed themselves with hand grenades.

That evening, news came of the Victory in Europe, which was great; but for us, we were still at war. That night our combat team was attacked by enemy infiltrators, the Japanese were expert night warriors attacking in stealth. Silently in the darkness of night, they would sneak in and slit the throats of our soldiers then secretly disappear.

Tank with an active flamethrower (The Army called a tank with a flamethrower a Blowtorch)

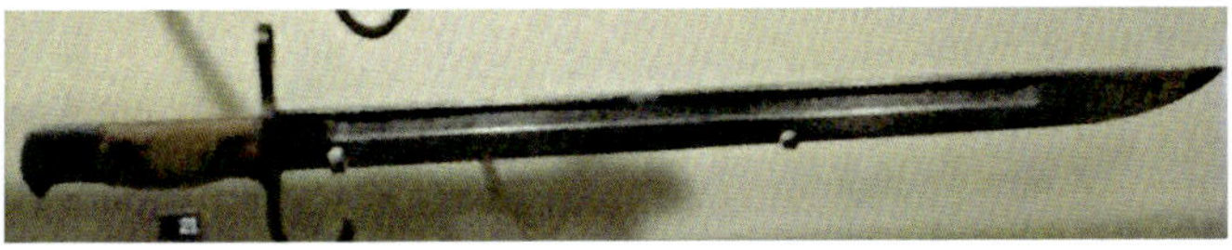

Type 30 Bayonet can be attached to a Type 38 and Type 99 Japanese rifle as well as the Type 99 Light Machine Gun.

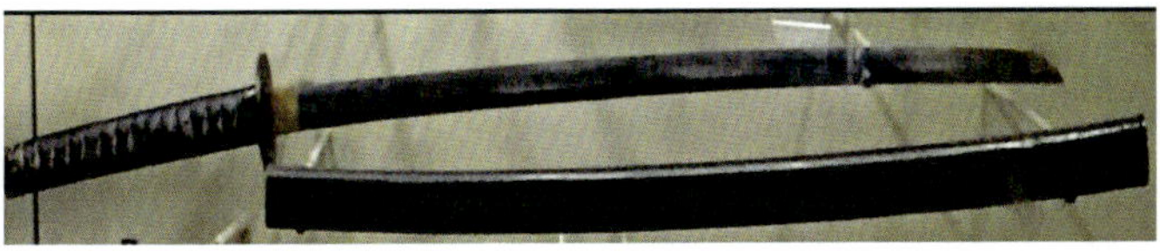

In keeping with the Japanese Samurai tradition, many military officers carried family swords that had been passed down for generations. To the Japanese, these swords were not carried as a symbolic badge of office but were wielded as fearsome weapons in combat.
Pictures: The National World War II Museum, New Orleans, Louisiana

The beginning of Night Vision in Warfare:

The Japanese were superior nighttime warriors, it was during The Battle of Okinawa that America began to develop and use a Carbine with an infrared scope similar to the one pictured below- the German Military began to develop night optics as early as 1939. Pictured below is the M3 Carbine with Infrared Scope used in the Korean War.

Picture: The National Infantry Museum, Fort Benning, Georgia

A 60 mm M2 Mortar: Picture from The National Infantry Museum, Fort Benning, Georgia

May 10th, 1945

On May 10th near Shuri Ridge, we were able to close several cave openings with heavy artillery while in a fierce battle. At the time, I was a Sergeant over a fifteen-man 60 mm M2 Mortar Platoon. After the battle, we dropped white phosphorus grenades into the tunnel openings used by snipers; it was something that had to be done. To be honest from my observation we were up against thousands of Japanese. The soldiers I admired most in these battles were our medics, they never left a soldier behind; always risking their lives to save our wounded, often under fire and sometimes wounded or killed in their mission. One of the enemy's strategies to deter our advance was to intentionally kill our medics, thinking it would stop our attack.

Picture :The National World War II Museum, New Orleans, Louisiana

Hand Grenades: The fragmentation grenade Mk II A1 (Pineapple) shown above weighs approximately 20 ounces and has enough bursting charge to send segments flying 200 yards. Other types of grenades include incendiary grenades, colored smoke grenades, illuminating grenades, offensive grenades, and chemical grenades which may be filled with different types of gasses.

Picture: The National World War II Museum, New Orleans, Louisiana
Top Weapon, M1928 Thompson Submachine Gun: Bottom Weapon, the M3 "Grease Gun" Submachine Gun
Both of these Submachine Guns were used in the Pacific Campaign in Okinawa

Picture: The National World War II Museum, New Orleans, Louisiana
M1917A1 .30 Caliber Browning Water Cooled Machine Gun

Picture: The National World War II Museum, New Orleans, Louisiana
M2A1 105MM Howitzer High Explosive (HE) semi-fixed ammunition, Range 7 miles

Picture: The National World War II Museum, New Orleans, Louisiana
155 MM Gun M1 Long Tom, Barrel length 22 feet, Range 14.7 miles

Picture and quote: The National World War II Museum, New Orleans, Louisiana
The M1 Garand Rifle, nine-pound semiautomatic, Caliber .30-06, ***"In my opinion, the M1 Rifle is the greatest battle implement ever devised." General George S. Patton***

Chapter Eight

Chocolate Drop Hill

Matthew 17:20 For verily I say unto you, if ye have Faith as a grain of mustard seed, ye shall say unto this mountain, Remove hence to yonder place, and it shall remove, and nothing shall be impossible unto you. The Bible (KJV)

It was named Chocolate Drop Hill by the soldiers of the 77th Infantry Division, our Headquarters called it Hill 130, and by the end of the battle everyone called it, "The Hill."

Picture: Chocolate Drop Hill: history.army.mil:Chapter XIII, The May Attack on The Shuri Defenses, page 340

As the 77th Infantry Division with a Marine Division moved south closing in on the Shuri Line of Japanese defense the fighting became more intense. Some historians believe the Imperial Japanese wanted to prolong the Battle of Okinawa for two reasons: The first: It would allow the Imperial Japanese more time to fortify the island of Japan defensively in anticipation of a final battle. The second: If the Allies especially the Americans experienced enough loss and pain on Okinawa an Armistice more in favor of the Imperial Japanese could be reached.

May 11th, 1945

The Hill

Sergeant Rathel, speaking

The first time I saw Chocolate Drop Hill was on May 11th, 1945, while in a battle against 200 Imperial Japanese near the top of a ravine. From my position, I could see the bursts of shells and hear the ripping thunder of tanks in a raging conflict. The earth vibrated and the explosive sounds were deafening as the artillery shells of the 305th and 306th Regimental Combat Teams led a charge up Chocolate Drop Hill under a barrage of enemy fire.

Later that evening radio communications revealed our soldiers were unable to rout the enemy. The slope around The Hill was covered with land mines and each time our soldiers tried to advance they were repelled by a blizzard of enemy fire. The position of "The Hill" allowed the enemy a perfect crossfire defense with machine guns placed on Flat Top Hill to the east, Ishimmi Ridge to the southwest, and smaller fortified knobs and mounds in between. These defensive locations made Chocolate Drop Hill almost invincible. Tanks firing 75mm shells knocked out a few machine guns but the enemy prevailed, three of our tanks were destroyed and six were damaged by enemy fire, our infantry was unable to advance and fifty-three of our soldiers were killed. As night arrived everyone knew t*his was going to be a major battle.*

Picture: The National Infantry Museum, Fort Benning, Georgia

The top shell is a 75mm M-61 APC, which was the primary armor-defeating round for the Medium Tanks. This round was capable of penetrating 3.7" of armor at 500 yards. The length of the round is 26.29 inches, weight is 19.92 pounds.
The bottom shell is a 37mm M-51 APC, which was the primary armor-defeating round for the Light Tanks. The round was capable of penetrating 1.8" of armor at 500 yards. The length of the round is approximately 12.81 inches, weight is approximately 2.62 pounds.

May 12th, 1945

During the night a decision was made to bomb and strafe enemy positions along the ridges near Chocolate Drop Hill and on highway 5 leading to Shuri. On the morning of May 12th, the Army Air Force strafed enemy ridge locations near Chocolate Drop Hill in waves with bombs and machine-gun fire while maintaining distance from our fighting teams.

Picture F4U-Corsair:https://www.history.navy.mil/content/history/museums/nmusn/explore/photography/aircraft-us/aircraft-usn-f/f4u-corsair/80-g-49320.html

May 13th, 1945

The Triangle of Death

In the valley of Chocolate Drop Hill, we found ourselves in a triangle of death, overwhelmed by a blizzard of enemy machine guns, mortar, and artillery fire. From elevated positions on The Hill, with crossfire locations on Wart Hill and Flat Top Hill, the enemy had the advantage. The battle raged!

Actual picture of a battle scene in Okinawa found in a donation of a deceased soldier's memories who served in the Battle of Okinawa: The Woodring Wall of Honor and Veterans Park, Enid, Oklahoma

May 14, 1945

On this fourth day of battle, a decision was made to combine two battalions of the 306th Combat Team into one composite battalion due to extremely high casualties. Because of the magnitude of the battle, the 307th was repositioned to replace the 306th on the left side of the hill. Our priority was to eliminate the two crossfire positions which would allow us to advance. An additional danger we constantly faced was snipers. The Japanese were masters of camouflage and to our eyes they were unseen; hiding in the cracks of ridges or the sides of hills sometimes appearing from openings in concealed tunnels behind us. They attacked with lightning speed-- killing our men instantly and then disappearing into the terrain.

Picture, The National Infantry Museum, Fort Benning, Georgia

May 15th, 1945

Holding The Hill

After an all-night bombardment of "The Hill" by our artillery, on the morning of May 15th, the 307th completely replaced the 306th Regimental Combat Team. The 306th had been in battle continuously for ten days and had experienced many casualties, it was time for them to rest and reorganize. Fortunately, the 305th and the 306th had broken through the Japanese front line of defense near the base of Chocolate Drop Hill. Our mission now was to gain ground and hold "The Hill." Our Army had already taken and lost Chocolate Drop Hill three times. As we moved forward in battle two battalions of the 305th attacked enemy crossfire positions on Flat Top Hill and Wart Hill.

May 16, - May 17th, 1945

Defining Courage

After two days of a fierce battle on the evening of May 17th, 1945, we were able to gain control of this 140-foot hill. For me, Chocolate Drop Hill will always be remembered as a place of Defining Courage. I say that in Honor of the courage exhibited by the soldiers who served, gave their lives, were wounded, and whose blood was sacrificed on, "The Hill." These Soldiers gave their all to the mission at hand, "they never surrendered, and even in the direst of circumstances they never left a Soldier behind, we were family we took care of our brothers in arms." These soldiers held the Torch of Liberty High, they were True Heroes, Men of Valor.

Captured Japanese Type 95 HA-GO Tank, Picture from the donated memories of a deceased soldier who served in the Battle of Okinawa: The Woodring Wall of Honor and Veterans Park, Inc. Enid, Oklahoma

Chapter Nine

Salient—The Lost Company of Ishimmi Ridge

May 17th to May 19th, 1945

While the final battles on Chocolate Drop Hill were in progress plans were being made for the attack on Ishimmi Ridge by the 307th Regimental Combat Team, the Company chosen for this Honor was Company E, with support from Companies F, G, H, and one platoon from Company C. Company E's appointment with destiny on Ishimmi Ridge would begin before dawn on the 17th of May and be recorded forever in military history. This historical event was reminiscent of the ***Famous Lost Battalion in the Argonne Forrest*** *when nine companies of The 77th Infantry Division and Company K of the 307th Infantry were isolated by the German Army during* ***World War I.***

Historically The Battle of Ishimmi Ridge will be remembered because the Soldiers involved demonstrated fearless *loyalty under fire when they became separated from their Division and found themselves surrounded in the very heart of the Imperial Japanese Army. Yet this battle in the end provided a Breakthrough for the 77th Infantry Division and was* ***The Key to breaking the Shuri Line of Defense leading to the ultimate Victory in Okinawa.***

—Important—

Of the 204 Soldiers who served on Ishimmi Ridge from May 17th to May, 19th, 156 became casualties, according to official military records. (*14)

Picture: The National World War II Museum, New Orleans, Louisiana

(*14) https://www.history.army.mil/books/wwii/okinawa/chapter13.htm, The May Attack on the Shuri Defenses Page 338 Paragraph 2

Chapter Ten

My Reflections on The Battle of Chocolate Drop Hill

May 18- May 19, 1945

Sergeant Rathel, speaking

The next two days were spent closing cave openings by blasting them with our tanks, flamethrowers, and dynamite; destroying enemy equipment including ammo, and taking two enemy prisoners we found hidden in the caves. These prisoners were injured and too weak to commit suicide. In this war once cornered the enemy committed suicide rather than surrender. While closing the caves we still faced sniper attacks from scattered enemy soldiers.

Captured Japanese Type 95 HA-GO Tank: Picture from the donated memories of a soldier who served in the Battle of Okinawa, The Woodring Wall of Honor and Veterans Park, Inc. Enid, Oklahoma

What we discovered after the Battle of Chocolate Drop Hill

Chocolate Drop Hill had been bombarded by thousands of shells from our Army and Navy. Our Tactical Air Force had dropped tons of bombs, showered it with Napalm, and machine-gunned its armed locations almost daily; still, Chocolate Drop Hill remained invincible. It was a fortress, after the battle we discovered, hundreds of cave openings on its reverse slope, these caves were so well fortified the Imperial Japanese could go deep underground when being bombed. The caves were tiered with levels having petitioned floors with each level containing weapons, heavy artillery, tanks, ammunition, and supplies; many of the tiered large rooms connected to an intricate network of tunnels with openings used for heavy artillery and machine gun pill boxes. These tunnels also provided openings for sniper holes in the ridges and also underground openings used for infiltration and stealth attacks. As we destroyed the caves with artillery there were enemy encounters from those that refused to admit defeat.

Battle Scars

The one thing I remember most about the battle of Chocolate Drop Hill is wounded soldiers bleeding and dying in the mud and rain. Sometimes, even though it was long ago, I awaken during the night hearing the death cries of wounded soldiers. I still have dreams of crawling under enemy fire trying to reach a wounded soldier. To this day certain smells remind me of the smell of burning flesh caused by the white phosphorus hand grenades we used against the enemy. That smell mixed with battle smoke is something I'll never forget. I guess you could call these emotional battle scars. Scars imprinted on my soul that will be with me the rest of my life

Sometimes when I'm alone flashes of battle scenes I've witnessed come to mind. One scene, in particular, was an incident that happened on "The Hill." In the battle a mortar exploded just a few yards in front of me, killing seven soldiers instantly; the visual memory of this tragedy is indescribable. These soldiers were my buddies and all I could do was grit my teeth and stay alert while under attack; putting my emotions on hold.

A Moral Dilemma, Killing the Enemy,

The one issue soldiers never talk about is a moral dilemma for many, "killing the enemy;" it's a debate hidden and locked deep within our souls for life. The issue is guilt, each soldier confronts it differently. The reality is as soldiers we are the last line of defense for our country, and sometimes it requires us to be on the defensive and kill the enemy, that's why we're called Soldiers.

Yes! I thought of those we killed, "the enemy." My first kill was the hardest, I tried to reason the issue away debating back and forth in my soul the big questions, Why? Why are we killing each other? How do I justify this? In Okinawa it was kill or be killed! In the end, I didn't feel guilt, but guilt is an emotion many of our soldiers toiled with. Sometimes guilt was not the problem; some soldiers become hardened, losing their human compassion and in anger kill the enemy -- especially if one of their buddies has been killed before their eyes. What made killing the enemy so hard in the Battle of Okinawa was the fact that the Japanese used children, young boys, called the "Iron and Blood Imperial Corps," It's true, they put them on the front lines in battle. They also used young girls as nurses, called the, "Lily Corps," to nurse their wounded.

The following is a statement from, "Understanding Himeyuri," school children of Okinawa, used in the Battle of Okinawa.

*"On the main island of Okinawa, there were 21 male and female secondary schools. As the battle began, students were mobilized from these schools. While female students worked mainly as nurses, male students were divided into upper-grade and lower-grade students. The former were organized into the "Blood and Iron Loyalist Students Corps," and the later into the Signal Corps." Of all the students in Okinawa at that time, as many as 2,000 perished on the battlefield."(*15)*

Did God make us for this?

I also remember the sound vibration of tanks, the loud explosions of heavy artillery, the flash of mortar attacks, the rain of steel from machine gunfire, the uncertainty of making my next step in fields of land mines, those are the things I remember about Chocolate Drop Hill. At times I thought, "could this be real? Did God make us for this?" Often on my knees in the bloody mud with tears in my eyes I found myself asking God for help, looking toward Heaven and praying, "Jesus please help me."

Frozen,

Due to the many casualties we were experiencing in Okinawa, many young soldiers were hurriedly trained in the United States, some just out of high school and immediately sent into battle on the front lines in Okinawa. Many of these soldiers were placed in my platoon and on more than one occasion I've witnessed a soldier freeze-unable to fire their weapon at the enemy. It's a deadly situation and it usually happened during an enemy charge; fortunately, we looked out for each other but sometimes it was too late. *(Phycologists have studied this, and it is called "Flight or Fight," which is a normal human response for some people when in immediate stressful danger.)*

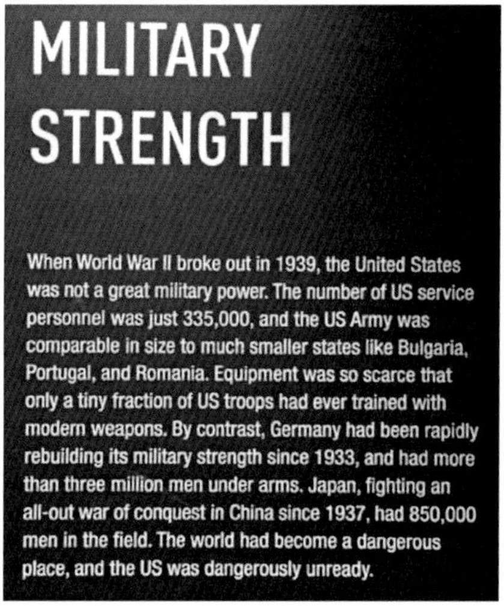

Pictures and statements of the Military Strength of the United States, Nazi Germany, and Imperial Japan at the beginning of World War II.

The National World War II Museum New Orleans, Louisiana

*(*15)"Understanding Himeyuri" Himeyuri Peace Museum 2006. Accessed December 27, 2022 https://www.himeyuri.or.jp/EN/war.html*

Shell Shock

Our soldiers also experienced the problem of shell shock and combat exhaustion from the constant loud explosions in battle. Shell Shock was a term used in World War I, to describe the effects of trauma on the body and mind, especially the nervous system. Near the end of World War II, the diagnosis was replaced with the term Combat Stress Reaction.

World War I Explanation of Shell Shock

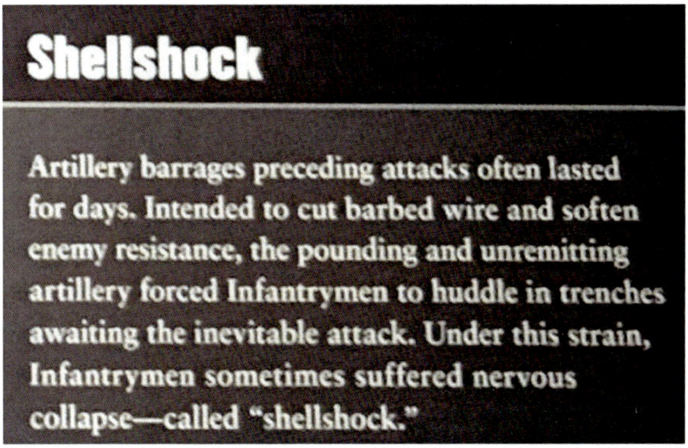

Statement : The National Infantry Museum, Fort Benning, Georgia

Chapter Eleven

Letters Covered with Sand

Often when a soldier was killed in battle there was a letter in his pocket addressed to his loved ones back home along with a small metal-covered New Testament Bible usually given to him by a caring family member or loved one.

Spiritual Warfare

Ephesians 6:10 Finally, my brethren, be strong in the Lord, and the power of his might. (NKJV)

Deuteronomy 4:39 Know therefore this day, and consider it in thine heart, that the LORD he is God in heaven above, and upon the earth beneath: there is none else. (NKJV)

Between battles, we often worshiped and prayed together.

Interfaith Chaplains,

"One of the most beloved members of any unit was often its Chaplains—Protestant, Catholic, and

Jewish—conducted worship services, provided spiritual and moral guidance, and worked to boost morale. These men of Faith, all volunteers, and commissioned officers, often accompanied frontline units and took great personal risks. Chaplin's offered comfort to the sick, wounded, and dying, and administered last rites when needed—often under fire. They also assisted with graves registration and conducted memorial services."

Pictures and statement: The National World War II Museum New Orleans, Louisiana

Chapter Twelve

Blue and Gold Star Families

Sergeant Rathel, speaking

Behind it all there was shared concern for our soldiers, from the leaders on the battlefield to our Generals behind the front lines. Everyone remembered these Soldiers had families who loved them, wives, children, and parents back home. With the high casualty rates, we were experiencing In Okinawa, that was a great responsibility.

Blue Star Families are those who have an immediate family member(s) serving in a time of conflict.

A Gold Star Family is the immediate family member(s) of a fallen service member who has died while serving in a time of conflict.

Gold and Blue Star Families may also display appropriate Gold or Blue Service Flags and wear appropriate Gold or Blue lapel pins in Honor of their family member(s).

Mother and Her Son, Gold Star Family

Pictures from The National Prisoner of War Museum, Andersonville, Georgia, The National Infantry Museum, Fort Benning, Georgia, and The National World War II Museum New Orleans, Louisiana

Blue Star Memorial, Butler, Georgia, Gold Star Memorial, Fort Mitchell National Cemetery, Fort Mitchell, Alabama

Chapter Thirteen

Deadly Friendly Fire

Picture: The National World War II Museum, New Orleans, Louisiana

May 21st, 1945

After the battle of Chocolate Drop Hill, the 77th Infantry Division moved forward penetrating the inner ring of the Shuri defense.

Before dawn on May 21st, 1945 my Company, "A" of the 307th Regimental Combat Team, prepared to attack a pocket of enemy resistance near, "The Three Sisters," (three hills forming a triangle just south of Flat Top Hill approaching Shuri). Our team consisted of about one hundred soldiers, eighty-eight of these soldiers were fresh replacements having no combat experience. As our team approached a gap between two hills suddenly, we were attacked. The attack came from riflemen of an adjoining regiment of the 96th Division. The attack alerted the enemy and a fierce battle en-sued. This was a well-documented case of Friendly Fire caused by miscommunication. As the battle ended several of our soldiers were wounded and one killed. Later investigation proved that the 96th Division had been notified of our battle plans the day before.

During the Battle, my right hand and arm were injured while in a bayonet confrontation defending a wounded soldier. I would rather die myself than see one of my soldiers killed. My hand and arm were cut needing medical attention.

Rathel, Clarence

RATHEL, Clarence, Pfc., son of Mrs. Lois H. Rathel, Gen. Del., [illegible]

Wounded in Action
Pacific Regions.

(Houston Co.)

Authority: OWI Casualty Release, June 21,1945.

Chapter Fourteen

Return of the Kamikaze

May 22nd, 1945

With my hand and arm injured I along with the other wounded on my team were taken for medical treatment. On the way, we experienced very heavy rainfall with flooding making some roads impassable. In the spring Okinawa receives the "plum rains," a period of heavy monsoonal rainfall. We eventually arrived at the harbor on the evening of the 22nd of May. We boarded a ship designated as a hospital ship. Once onboard, I could not believe how many critically wounded and dying soldiers there were. Hospital ships are supposed to be off limits to enemy attacks according to The Geneva Conventions but in this war the kamikaze were desperate. In fact, just the month before on April 28th The USS Comfort, a hospital ship, fifty miles off the coast of Okinawa had been attacked by the kamikaze.

While I was having my hand and arm examined the kamikaze started attacking ships in the harbor, suddenly alarms sounded as we prepared for a possible attack.

(*16) " Japanese air raids reached a peak during the latter part of May. On the 20th, thirty-five planes raided the American fleet, twenty-three were shot down. On 22 and 23 May, Japanese planes came over Okinawa again. Beginning on 24th was perfect bombing weather with a clear sky and full moon. The air alerts started about 2000 and it was 2400 before an all-clear sounded. In that interval, there were seven distinct air raids on Okinawa."
(*16) https://history.army.mil/books/wwii/Okinawa/ Chapter XIV: Battle in the Rain, page 361, paragraph 2.

Japanese Kamikaze; Aircraft— Mitsubishi Zero A6M5 model 52c

Picture: USS Alabama Battleship Memorial Park, Mobile, Alabama

May 25, 1945

Due to the kamikaze attacks, our ship received injured soldiers and sailors, on May 24th, many were wounded with some severely burned. On the morning of May 25th, I was awakened by a surprise, our ship began to sail. Concerned about the movement and worried about being missed by my Regiment, I asked the Doctor in charge what I should do, he informed me not to worry he would handle the notification properly and do all the paperwork. I then asked him, "Where are we going?" He replied, "the Mariana Islands." I choked while repeating the words, the Mariana Islands!

With my hand bandaged and my arm in a sling, I was allowed on deck as our ship moved out of the harbor. My eyes were weary of seeing the blood of war, I now focused on the beautiful seagulls following our ship's waves. Somehow at that moment, God spoke to my spirit as I watched a particular seagull fly higher than the others so oblivious to the sights and sounds of war. Then under my breath, I began to Praise God and pray for the men of my regiment. **Somehow deep within my soul, I began to release the pain, sorrow, and trauma of Okinawa;** wondering for the first time in a long time what the future might be. Once out of the harbor we passed a mighty battleship **the USS Mississippi as it sailed toward the coast of Okinawa, on a mission.**

On May 25th, 1945, the Battleship USS Mississippi (BB-41) began shelling Shuri Castle for three continuous days, until it burned. The Imperial Japanese Headquarters had evacuated Shuri Castle days before the bombing and had secretly moved to a location near the southern end of Okinawa, preparing for their last stand.

Picture: USS Mississippi (BB-41)

File:USS Mississippi (BB-41), USS West Virginia (BB-48) and HMAS Shropshire (73) bombarding Luzon on 8 January 1945.jpg. (2022, January 5). Wikimedia Commons, the free media repository. Retrieved 20:46, December 26, 2022 from https://commons.wikimedia.org/w/index.php?title=File:USS_Mississippi_(BB-41),_USS_West_Virginia_(BB-48)_and_HMAS_Shropshire_(73)_bombarding_Luzon_on_8_January_1945.jpg&oldid=619418532.

Finally, under relentless assault by the Tenth Army on May 29th Shuri Castle fell, then Marines seized the airfield at Naha through an amphibious assault commencing June 4th, 1945.

Picture: Shuri Castle in 1934

Shuri Castle in 1934.jpg. (2022, June 11). Wikimedia Commons, the free media repository. Retrieved 23:04, December 23, 2022 from https://commons.wikimedia.org/w/index.php?title=File:Shuri_Castle_in_1934.jpg&oldid=663675801.

Braving Japanese sniper fire, US Marine Lieutenant Colonel R.P. Ross, Jr. places an American flag on a parapet of Shuri Castle on May 29, 1945.

File:US Flag raised over Shuri castle on Okinawa.jpg. (2022, October 22). Wikimedia Commons, the free media repository. Retrieved 18:41, December 23, 2022 from https://commons.wikimedia.org/w/index.php?title=File:US_Flag_raised_over_Shuri_castle_on_Okinawa.jpg&oldid=698205360.

Chapter Fifteen

The Approaching End

Sergeant Rathel's story must be told in the light of world history and from this point forward the author will begin to set the stage for the most panoramic scenes of tragedy and triumph in the Twentieth Century.

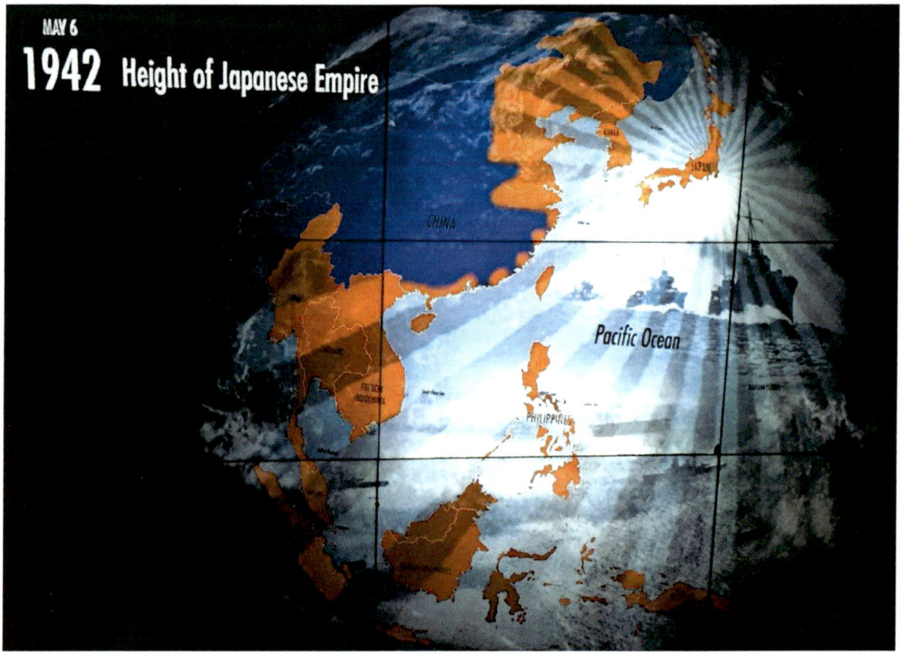

Revelation 6:8 And I looked, and behold a pale horse: and his name that sat on him was Death, and Hell followed with him. And power was given unto them over the fourth part of the earth, to kill with sword, and with hunger, and with death, and with the beasts of the earth. (The Bible NKJV)

Map: The National World War II Museum, New Orleans, Louisiana

Emperor Hirohito, The God Emperor of Imperial Japan allied with NAZI Germany, and Fascist Italy; approved military plans to establish a Japanese Empire in Asia and the Pacific. His unrelenting quest for power would soon place Japan in a "Fiery Furnace," and his choices would soon change world history.

HIROHITO

Japanese Emperor Hirohito ascended to the throne in 1926, assuming the role of God Emperor. He presided over a period of internal tumult, violence, and aggressive expansion. Allied with Nazi Germany and Fascist Italy, Hirohito did not object as Japan's military invaded China, and prepared for war with the United States. In November 1941, Hirohito personally approved military plans for a sweeping assault on the Allied powers to establish an empire in Asia and the Pacific.

A Letter That Changed History

The Einstein-Szilard Letter, August 2nd, 1939

One of the most important letters written in the twentieth century was a letter signed by Jewish Physicist Albert Einstein dated August 2nd, 1939 to President Franklin D. Roosevelt. The letter called "The Einstein-Szilard Letter" warned President Roosevelt of NAZI Germany's research into the power of atomic energy, and the possibility of Atomic Warfare in the future... The letter prompted our President and government to secretly create, "The Manhattan Project."

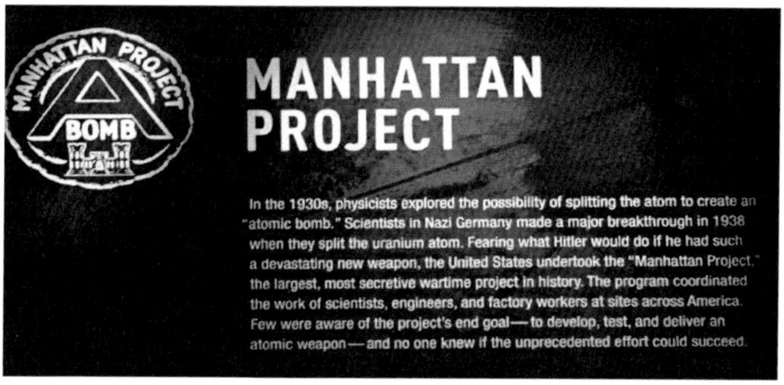

(Please notice the emblem with all of its words and the letter A)

Pictures and Statement: The National World War II Museum, New Orleans, Louisiana

The USS Indianapolis

On July 16, 1945, within hours of the Trinity test, the USS Indianapolis (CA-35) a Portland-class heavy cruiser on a Top-Secret Mission departed Hunters Point Naval Shipyard San Francisco, California, to deliver uranium and other components of the first Atomic Bomb used in war, "Little Boy," to Tinian Island a B-29 Superfortress Base in the Mariana Islands. Tinian Island's code name was "Destination."
(The Trinity nuclear test– see Chapter Twenty, "Trinity the Beginning of a New Age")

Important Note

After leaving Tinian Island the USS Indianapolis would be hit by two torpedoes from a Japanese submarine; the resulting tragedy would become the worst disaster in United States Naval History. The USS Indianapolis had a crew of 1195 men and only 316 men survived this horrific tragedy.

One can only imagine the outcome of World History had this ship been torpedoed en route to Tinian Island.

Picture: The USS Alabama Battleship Memorial Park, Mobile, Alabama

Picture: USS Indianapolis (CA-35) at Pearl Harbor, Woodring Wall of Honor and Veterans Park, Inc. Enid, Oklahoma

The B-29 Superfortress introduced by The Boeing Company on May 8th, 1944, would play a vital role in the war against Imperial Japan. The Superfortress was designed for high-altitude strategic bombing and low-altitude night incendiary bombing. B-29s would ultimately be used to drop atomic bombs on Imperial Japan.

Pictured above is the "Enola Gay," B-29 Superfortress, this aircraft dropped the first atomic bomb: The National World War II Museum, New Orleans, Louisiana (see Chapter Twenty Two, "My God What Have We Done")

Chapter Sixteen

June 5th, 1945

Conversations and Thoughts En Route to Tinian Island

Sergeant Rathel, speaking

From Okinawa to Tinian Island is a long way when the world is at war. Somebody said as the crow flies it's only 1,500 miles. But it's different when you are on a ship making stops to pick up injured sailors, soldiers, and Marines, and dodge enemy aircraft, and submarines. I can't remember how long it took for sure, but I think it was close to three weeks. En route I tried to keep up with the news about the battles in Okinawa, I also tried to find out what was going on in the States. The other thing I did was enjoy real food, the Navy fed us hot meals and after almost starving in the battles on Okinawa, real food was a luxury. I also enjoyed showers and being clean; in Okinawa, the mud was ever-present and you never felt clean. I also consoled and helped the wounded and sick on board, I still could use my left hand.

A few days after we left Okinawa news came that after being pounded by thousands of rounds from our Navy and being bombed by our Tactical Air Force, beautiful Shuri Castle had collapsed and burned. I later heard that an amazing discovery was made at the site. Two large Ancient Bronze Bells had been uncovered in the rubble, one was five feet high and the other about three and a half feet high. The last three lines inscribed on the Bells after translation had an ancient spiritual message from the Ryukyu Kingdom of 1550.

"And how will the bell sound? It will echo far and wide like A Peal of Thunder, but with utmost purity. And evil men, hearing the bell will be saved."(*16)

Shuri Castle Bell, with American Officer standing by. Bell is a companion to one brought to The U.S. Naval Academy by Commodore Perry.

*(*16) https://history.army.mil/books/wwii/okinawa/chapter15.htm The Fall of Shuri: picture page 401, statement page 402 paragraph*

One of the things I've always admired about the Navy is the respect they give each other including their passengers and patients. As we sailed everyone enjoyed big band music, I guess it was popular back home. We also would gather around the Steinway Piano and sing hymns on Sundays after the Chaplin gave Communion.

Pictured are ***examples of items*** *from The USS Alabama Battleship Memorial Park, Mobile, Alabama*

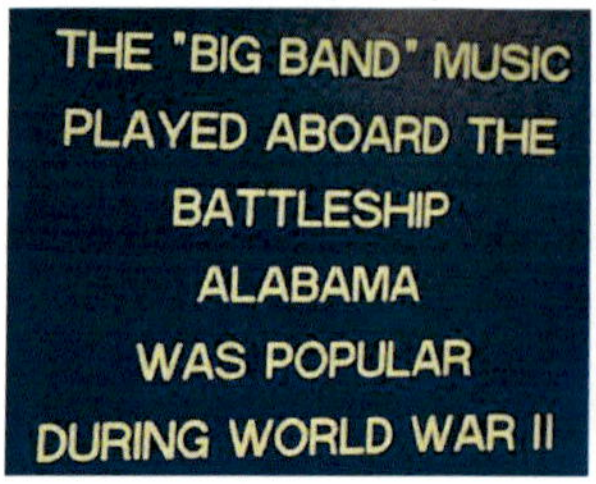

Portable Communion Articles

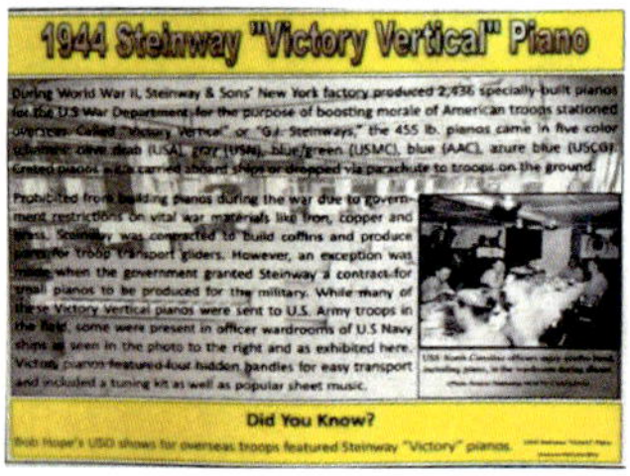

Steinway Piano and Hymnal

While on board the ship I was fortunate to meet several very interesting people and of course, we were able to talk about life back home in the states. A few of the patient soldiers had just arrived in Okinawa when injured in their first battle. These soldiers had current news about home and I was eager to hear what they had to say. I had lots of questions.

Japanese Americans,

In one conversation I asked how the Japanese Americans were being treated back home; their answers astonished me. I had no hatred for the Japanese and couldn't believe the Liberty we professed as Americans didn't apply to all of our citizens. The fact that our country had placed 110,000 Japanese Americans in incarceration camps, during the war got my attention.

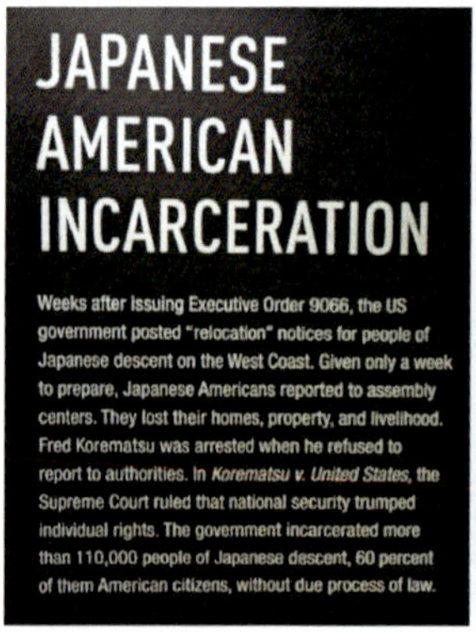

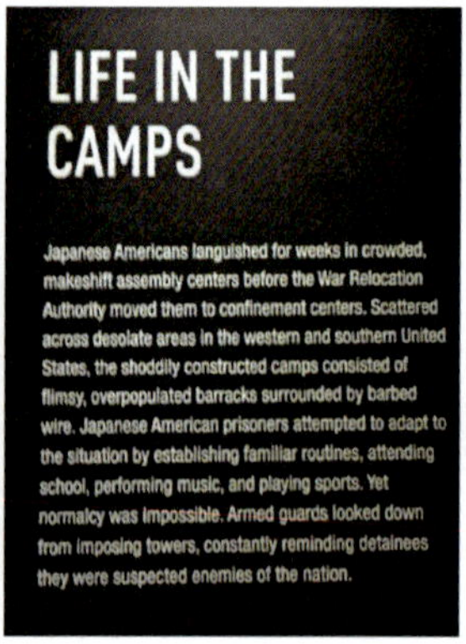

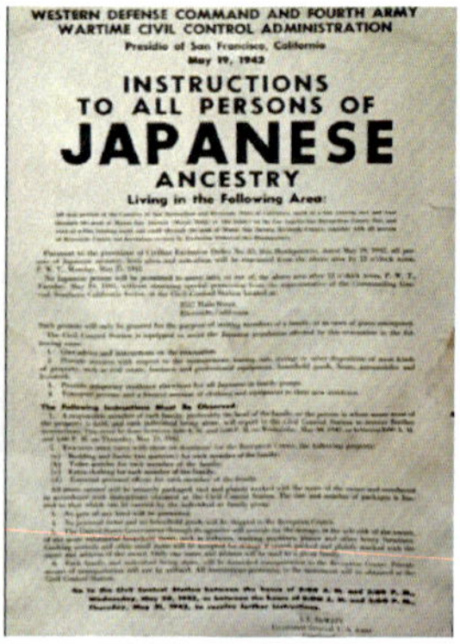

Statements: The National World War II Museum, New Orleans, Louisiana

Native American Indians,

One of the patients I became friends with was a Native American Indian. I was curious and asked him how Native American Indians were affected by the war. He said that he and many of his friends joined the Army and considered it an honor to serve.

"Despite a history of mistreatment by the US government, Native American tribal councils declared support for the war effort. After Pearl Harbor, Native Americans around the country enlisted to defend the United States. More than 70 percent of the men in some Native American communities joined the military during the war. Many learned basic English to meet enlistment requirements. Fighting in integrated units and demonstrating tremendous skill and bravery. Native Americans earned the respect of their fellow soldiers."

Statement and picture: The National World War II Museum, New Orleans, Louisiana

Code Talkers: Soldiers from at least fifteen Native American Indian Nations used their traditional tribal language in World War II as Code Talkers to win victory for the Allies in both the European and Pacific Campaigns of World War II.

Women Air Force Service Pilots

One day in a conversation with a nurse, I was told that back home some of the WASPs (Women Air Force Service Pilots) were piloting B-17s for the military even though they were civilians, I guess everybody was helping win the war. *(These pilots had a passion to serve our Nation, some even losing their lives in aircraft accidents.)*

These four pilots leaving their ship at the four-engine school at Lockbourne AAB Columbus, Ohio are members of a group of WASPs who have been trained to ferry the B-17 Flying Fortress.
Picture: The USS Alabama Battleship Memorial Park, Mobile, Alabama

Alabama Heroes

By the end of the war, the Tuskegee Airmen earned respect and acknowledgment for their vital contribution towards the Allied victory in Europe. Over 100 enemy planes were destroyed by the airmen over the course of the war. Military medals for bravery awarded to Tuskegee Airmen include: 1 Legion of Merit, 1 Silver Star, 2 Soldier Medals, 150 Distinguished Flying Crosses, 8 Purple Hearts, 14 Bronze Stars, and 744 Air Medals. Their success, in the face of many obstacles, continues to inspire us to this day.

The Tuskegee Airmen

"Red Tail Angels"

One wounded soldier from Alabama told me about some amazing war heroes from our home state who were pilots in the Mediterranean and European Campaigns called The Tuskegee Airmen. I remember smiling with pride when I heard this.

Pictures and statements: The USS Alabama Battleship Memorial Park, Mobile, Ala

Kanti Kessen, Japanese Naval Battle Strategy

One of the most interesting conversations I had while on my journey to the Marianas was with a high-ranking sailor who had survived the Pearl Harbor attack. I asked him who designed the Pearl Harbor attack and where the Japanese doctrine of sea warfare came from. He explained that Imperial Japanese Admiral Isorouko Yamamoto who had attended Harvard in the United States planned the Pearl Harbor attack. He also explained that the Japanese Naval Battle strategy called "Kantai Kessen," was designed by American Alfred Thayer Mahan a United States Naval Officer and Historian of the nineteenth century, who wrote books about Naval War strategy.

Admiral Isoroku Yamamoto

Isoroku Yamamoto was born on 4 April 1884. He attended the Imperial Japanese Naval Academy and then Harvard University in the United States. Yamamoto eventually rose to be the Japanese Marshal Admiral of the Imperial Japanese Navy and the commander-in-chief of the Combined Fleet. He held these titles until his death. He played a large part in advancing Japanese aviation and even opposed the building of the super battleships Yamato and Musashi. While he thought war with America and Europe was a bad idea he thought it inevitable. He planned a dangerous raid on Pearl Harbor, Hawai'i with the intention of crippling the United States Navy's Pacific Fleet and hinged on Japan's six fleet carriers. His flagship for the attack was the battleship Nagato.

Kantai Kessen

In May 1905, on the Sea of Japan, at the Battle of Tsushima, the Decisive Battle Doctrine cemented its place in Japanese Naval strategy. Called *Kantai Kessen* (艦隊決戦) by the Japanese, this strategy relied on one large decisive naval battle to win a war. This theory was originally put forward by American author and naval historian Alfred Thayer Mahan in a number of his books. This tactic paid off for the Japanese during the Russo-Japanese War at the Battle of Tsushima. It was here that large naval rifles with long range and a single decisive battle could end a conflict and allow a smaller force to defeat a larger one. The Imperial Navy took this to heart and this *Kantai Kessen* would be the Imperial Navy's doctrine through the Second World War. At first it hinged on those large caliber long range guns, but as time went on a debate began to rage concerning air power. Many in the Imperial Navy staunchly laid in the "big gun" camp, while others, like Admiral Isoroku Yamamoto, believed carriers and air power would rule naval conflict in the 20th century. While the doctrine remained the same, relying on a single decisive battle, the Imperial Navy looked more and more at carriers and their ability to project power. As naval power evolved the site of the proposed battle moved farther and farther from the home islands of Japan.

Pictures and articles: The USS Alabama Battleship Memorial Park, Mobile, Alabama

Ambassador-Soldiers in a foreign land,

June 18, 1945

As we sailed toward the Mariana's, I found myself gazing across the endless horizon of the Pacific Ocean and for the first time in three years there was a calmness in my spirit. After the tsunami of death I had experienced in Okinawa it was time to praise and reconcile with God. In prayer I thanked Him for allowing me to survive the battles in Guam, Leyte, Ie Shima, the Kerama slands, and Okinawa and yielded to the fact that He had chosen me to live in this most turbulent time in history. As I considered the present I wondered would there be another battle, or possibly even another war in my future. Through all the destruction I had witnessed I knew The Lord had been with me and I continually held onto His promise that He would never leave me or forsake me. (Deuteronomy 31:8) As a soldier my focus and attitude was always on the mission, and because our country was still at war, I wondered what it would be. It was during these thoughts The Lord reminded me that as Christians we are His ambassadors in the world, and our home is really in Heaven. While In that train of thought I remembered the first time the people of Okinawa saw our soldiers marching through their village, mothers with their infants and children running away afraid. We were probably the first Americans they had ever seen and I guess the Japanese had filled them with fear. I know to them we were seen as the invaders. Maybe we looked strange in their eyes, being a combination of different races, and I'm certain they were curious. I only wished they had known the purpose of our mission was for their liberty. In the end I prayed that after the war they would know America was their friend.

This photograph was found in the collection of memories of a soldier who served in the Battle of Okinawa, donated to The Woodring Wall of Honor and Veterans Park, Inc. Enid, Oklahoma.

Chapter Seventeen

Okinawa The Cost in Blood

June 22nd, 1945

"In Flanders Fields the Poppies blow, between the Crosses, row on row..."

Sergeant Rathel, speaking

On the 22nd of June, our ship arrived at Tinian Island in the Marianas, as we docked a B-29 Super-fortress was taking off on a bombing mission. While we were still on board an announcement was made stating our country had Liberated Okinawa.

Once on shore, everyone was excited. I was joyous, then suddenly my eyes filled with tears as memories of the bloody battles I had encountered in Okinawa filled my soul with bittersweet emotion. Just as quickly I returned to the present waving my good hand trying to conceal my true feelings...

Okinawa The Cost

"The price paid for Okinawa was dear. The final toll of American casualties was the highest experienced in any campaign against the Japanese. Total American battle casualties were 49,151 of which 12,520 were killed or missing and 36,631 were wounded. Army losses were 4,582 killed 93 missing, and 18,000 wounded; Marine losses, including those of the Tactical Air Force, were 2,938 killed and missing and 13,708 wounded; Navy casualties totaled 4,907 killed and missing and 4,824 wounded. Non-battle casualties during the campaign amounted to 15,613 for the Army and 10,598 for the Marines.

*The cost of the battle to the Japanese was even higher than to the Americans. Approximately 110,000 of the enemy lost their lives in the attempt to hold Okinawa, and 7,400 more were taken, as prisoners." (*17)*

According to an article titled, "Terrible Losses," at the National World War II Museum in New Orleans, Louisiana, the deaths of the citizens of Okinawa may have been 150,000. (From the author, the question is, Why?)

*(*17)https:// history.army.mil/books/wwii/okinawa/chapter18.htm The Battle Ends page 473 paragraph 2 and page 474 paragraph 1*

"To you from failing hands we throw, The Torch; be yours to hold it High"

7Th Division Cemetery, Okinawa, 1945, this picture was discovered in a small suitcase containing a soldiers memories of The Battle of Okinawa, it was recently donated to the Woodring Wall of Honor and Veterans Park, Inc. Enid, Oklahoma.

Requiem

June 23, "Okinawa Memorial Day"

In Japan's Okinawa Prefecture, "Okinawa Memorial Day" is a public Holiday observed on the 23rd of June each year, its official name is, "Irei no Hi," "the day to console the dead." It is observed to honor and remember those who lost their lives (military and civilian) in "The Battle of Okinawa."

On this date in 1995, the fiftieth anniversary of the end of the Battle of Okinawa and in memory of the end of World War II, a special monument, "The Cornerstone of Peace," was unveiled in Itoman City. Inscribed on the monument are the over two hundred forty thousand names of those who perished in the battle. This monument is a place of memory, prayer, and hope for world peace.

Okinawa Prefectural Peace Park

File:Okinawa Heiwakinen Memorial Park01n3104.jpg. (2022, March 28). Wikimedia Commons, the free media repository. Retrieved 22:27, January 5, 2023 from https://commons.wikimedia.org/w/index.php?title=File:Okinawa_Heiwakinen_Memorial_Park01n3104.jpg&oldid=644723375.

Chapter Eighteen

The Sun Begins to Set on Imperial Japan

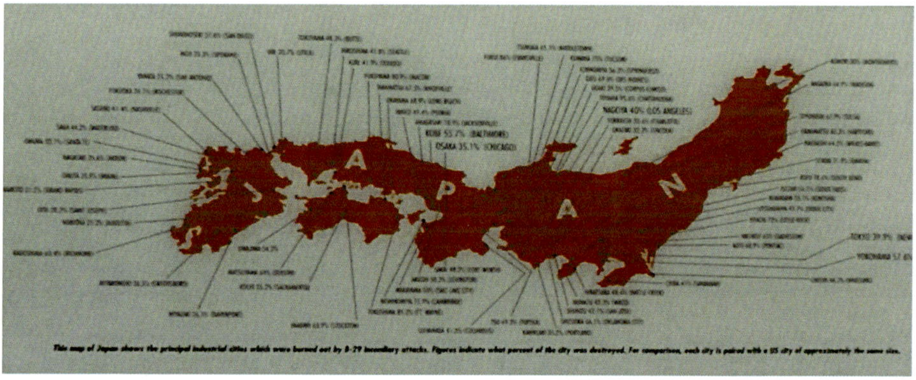

Picture: The National World War II Museum, New Orleans, Louisiana

This map of Japan shows the principal industrial cities which were bombed by B-29 incendiary attacks starting on March 29, 1945, and lasting till August 1945.

Sergeant Rathel, speaking

July 4th, 1945

By July 4th, my hand and arm had improved and I returned to light duty. I soon knew the B-29 bombing missions from Tinian to Japan were a twenty-four-hour operation. There were at least 250 B-29's based at Tinian airport which had four long runways and operated continuous bombing missions to Japan. Tinian was also a large port for the U.S. Navy and the talk was our submarines were destroying Japanese shipping.

The following statements and picture: The National World War II Museum, New Orleans, Louisiana

"Bombing Japanese Cities"

"Adverse weather over Japan plagued early B-29 missions, which were carried out at high altitudes and in daylight. On March 9, 1945, Major General Curtis LeMay radically changed tactics and ordered
a daring, low-level, nighttime strike on Tokyo using incendiary rather than conventional bombs. The attack ignited a firestorm that consumed 16 square miles and killed more than 100,000 people. More firebombing and daylight precision raids followed. By August 1945 nearly every targeted Japanese city
lay in ruins, and the nation's economy neared collapse."

B-29's dropping Bombs on Japan summer 1945

"Japan faced a blockade and relentless air attacks by the summer of 1945. Prowling submarines and offshore minefields effectually cut off foreign sources of food, oil, and other natural resources. B-29s continued ravaging Japanese industry, firebombing city after city. Aircraft from U.S. Navy aircraft carriers and newly secured bases at Iwo Jima and Okinawa struck at military and transportation targets, severing logistical links with Japan itself. The Allies' Naval dominance allowed closer maneuvering by American and British battleships and they soon joined the bombardment. Japan was reduced to desperate circumstances- yet would not surrender."

Chapter Nineteen

Operation Downfall, The Battle That Never Happened

July 1945

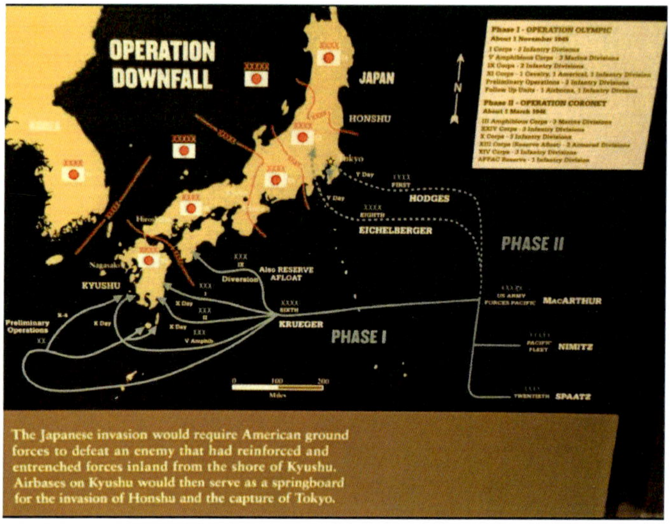

Map: The National Infantry Museum, Fort Benning, Georgia

As the Battle of Okinawa ended plans were being made by the Allies to end the war with Japan by conventional warfare through invasion. As these plans were being formed—only President Truman and those in the Top Secret, "Manhattan Project," were aware of the development of atomic bombs. Once atomic bombs became available General George Marshall of the United States Army was advised. Atomic bombs were originally going to be used to support the invasion forces of Operation Downfall. Consideration was also given to possibly dropping multiple atomic bombs before the invasion.

(Special note from the author: "Operation Downfall," is one of the most interesting topics I've discovered while researching the events leading to the end of World War II. There were many different strategies and opinions on how to end the war from the top leadership of the Allies, especially the Americans. The Imperial Japanese also had plans that will astound any reader of World War II history.)

Chapter Twenty

Trinity, the Beginning of a New Age

July 16, 1945

The photograph below is of the first atomic bomb explosion test, codenamed "Trinity," conducted by the United States Army, as part of The Manhattan Project. The explosion occurred at 5:29 am on July 16th, 1945, in the desert of New Mexico on the United States Army Air Force Alamogordo Bombing and Gunnery Range.

Picture: The Trinity Test Fireball

Description: PA-98-0520- Trinity Site explosion, 0.016 second after explosion, July 16, 1945
Date: 16 July 1945
Source: HTTP://www.lanl.gov/orgs/pa/photos/images/PA-98-0520.jpeg
Author: Berlyn Brixner / Los Alamos National Laboratory
Permission: This work is in the public domain it is a work prepared by an officer or employee of the United States Government as part of that person's official duties under the terms of Title 17, Chapter 1, Section 105 of the US Code.

Chapter Twenty-One

The Potsdam Declaration

July 26, 1945

Soon after Germany was defeated in Europe the Allied leaders, President of The United States Harry Truman, British Prime Minister Winston Churchill, and Chinese Nationalist leader Chiang Kai-sheik met in Potsdam, Germany on July 26th, and drafted The Potsdam Declaration, terms for Japan's surrender. Japan gave no reply to the Declaration which also included an ultimatum to disarm. When Japan dismissed the terms of surrender the Allies responded quickly.

On July 28th Clement Attlee was elected as the new Prime Minister of Britain, he was also present at the Potsdam Conference. General Secretary of the Soviet Union Joseph Stalin was present, but because the Soviet Union was not at war with Japan the Soviet Union was not part of the Declaration. On August 8th, the Soviet Union declared war on Japan.

Statement: The National World War II Museum, New Orleans, Louisiana

Chapter Twenty-Two

"My God What Have We Done"

August 6th, 1945

The following statements and pictures: The National World War II Museum, New Orleans, Louisiana

Hiroshima, The First Atomic Bomb

"On August 6, 1945, Captain Robert A. Lewis co-pilot of the "Enola Gay" kept a log of the atomic bomb mission as he flew out from Tinian to Hiroshima, then back. In the log, Lewis describes the mushroom cloud rising to an altitude of over 30,000 feet in less than three minutes. After his description of the blast, Lewis writes, "My God what have we done." The log stayed in Lewis's possession until 1971, but before parting with the original manuscript, Lewis hand-copied six exact replicas of the document and gave one to his wife and five children. This copy was a gift from Lewis to his youngest son, Steven. The original log sold at auction in 2002 to an anonymous buyer, and since then has been displayed publicly."

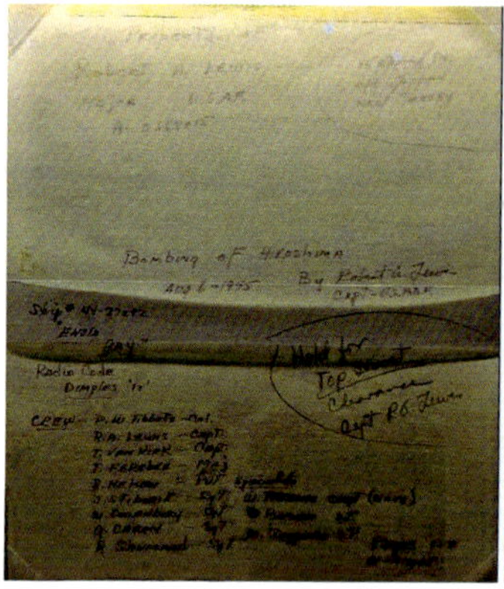

Captain Lewis's log August 9, 1945, Actual Photograph of the Hiroshima bombing

"This amateur photograph was taken from the town of Yoshiura, three minutes after the bombing of Hiroshima, Captain Robert A. Lewis acquired this photograph and wrote the caption at the bottom."
The caption: "Hiroshima 3 min. after detonation 9:18 AM August 6, 1945"

Nuclear weapon of the 'Little Boy' type the kind detonated over Hiroshima, Japan, in World War II.

Nuclear weapon of the 'Fat Man' type, the kind detonated over Nagasaki, Japan, in World War II.

The caption on the left reads Nuclear weapon of the "Little Boy" type, the kind detonated over Hiroshima, Japan, in World War II.
The caption on the right reads Nuclear weapon of the "Fat Man" type, the kind detonated over Nagasaki, Japan, in World War II.

"When Japan dismissed Allied surrender terms announced at the Potsdam Conference, the Allies reacted swiftly. On August 6, 1945, a rare clear day over the target, Colonel Paul Tibbets and the crew of the "Enola Gay" delivered the world's first atomic bomb, striking the city of Hiroshima. The bomb decimated the city, ultimately killing approximately 140,000 and wounding many more. When no surrender was forthcoming, on August 9, 1945, the crew of the B-29 "Bockscar" delivered a second atomic bomb. When their assigned primary target, Kokura, was obscured by clouds, they bombed their secondary target, Nagasaki, killing approximately 70,000."

The bombing of Nagasaki The B-29 Bockscar that dropped the bomb "Fat Man" on Nagasaki

Colonel Paul W. Tibbets, Jr.
The pilot of the B-29, The Enola Gay

The Flight Crew of the B-29, Bockscar

Destruction of Hiroshima Japan after the first Atomic Bomb attack: Picture: The National Infantry Museum, Fort Benning, Georgia

From the Author,

The dropping of Atomic Bombs on Japan resulted in an estimated 210,000 deaths and untold thousands wounded and injured for life from the effects of Radiation. Much has been debated in history and will continue to be debated about the decision to use nuclear weapons on Japan.

The atomic bomb dropped on Nagasaki was much more powerful than the bomb dropped on Hiroshima. The destructive effects of radiation caused by these two bombs continue to be studied by Biologists, Genealogist, Environmentalists, and many other fields of science.

In 1946 the United States conducted, "Operation Crossroads," a pair of nuclear tests using Atomic Bombs. These tests were conducted in the Bikini Atoll and were the first of many tests that continued into the 1950s in the Marshall Islands. The purpose of these tests was to determine the effects of nuclear weapons. As of this writing, January 2023, the Bikini Atoll remains uninhabitable due to radioactive contamination.

Chapter Twenty-Three

The Surrender of Japan

August 15, 1945

On August 8th, 1945 Soviet Russia declared war on Japan, this event along with the bombings of Hiroshima and Nagasaki, compelled Emperor Hirohito to announce the surrender of Japan on August 15th, 1945.

The following statement is from The National World War II Museum, New Orleans, Louisiana.

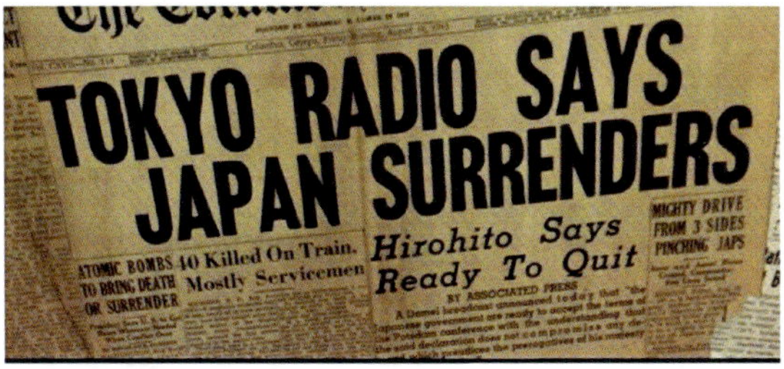

TOKYO RADIO SAYS JAPAN SURRENDERS

ATOMIC BOMBS TO BRING DEATH OR SURRENDER

40 Killed On Train, Mostly Servicemen

Hirohito Says Ready To Quit

MIGHTY DRIVE FROM 3 SIDES PINCHING JAPS

"The twin shocks of the Soviet Declaration of War and the Nagasaki bombing ignited fierce debates in Tokyo. Emperor Hirohito decided that Japan must "endure the unendurable." Although the Allied leaders rejected Japan's initial demand that the Emperor retain supreme authority during the occupation, Hirohito ordered his government to surrender. After outbreaks of rebellion. Japan's armed forces reluctantly complied. On August 15th, a radio broadcast of a recording of Hirohito's voice notified the stunned Japanese people that the war had ended. Hirohito would remain a symbolic figure as emperor. Japan signed the formal surrender documents aboard the battleship the USS Missouri (BB-63) in Tokyo Bay on September 2nd, 1945. World War II was over."

LEDGER EXTRA

JAPAN ACCEPTS PEACE TERMS

Newspaper Headlines Dated August 10, 1945: The Columbus Museum, Columbus, Georgia

Newspaper Headlines Dated August 14, 1945: The Columbus Museum, Columbus, Georgia

(Due to the International Dateline there is a time difference between Japan and the United States)

On August 14, 1945, America began celebrating V-J Day, Victory over Japan Day, the United Kingdom declared V-J Day as August 15. The official V-J Day is September 2nd, 1945, the day the formal surrender documents were signed.

Chapter Twenty-Four

Triumph and Surrender, the End of Global War

General Douglas McArthur accepted the unconditional surrender of Japan on September 2, 1945, aboard the battleship USS Missouri in Tokyo Bay.

Picture: The National World War II Museum, New Orleans, Louisiana

September 2nd, 1945, The Signing of the "The Instrument of Surrender" by Japan

At 9 am on September 2nd, 1945 one of the most monumental ceremonies in World History took place, the ending of a Global War- World War II. As this panoramic ceremony began a massive armada of over 300 ships was anchored in Tokyo Bay representing many of the nations of the world. On this triumphant day every detail of the ceremony was carefully choreographed with symbolism. Visible on the surrender deck in a frame was a special American flag, the Flag of the USS Powhatan, Flagship of Commodore Matthew Perry's expedition to open Japan to the world in 1854. The vessel chosen for this majestic ceremony and meant to symbolize the power of the United States was the Battleship USS Missouri (BB-63) named after President Harry Truman's home state and christened by his daughter Margaret upon being commissioned in 1944.

In anticipation of this historic event, thousands of sailors watched every movement on the USS Missouri, many from their vessels in Tokyo Bay. As the historic hour approached, military representatives from around the world gathered under the big guns of the Missouri, along with the Generals who had commanded the Allied Forces in war. In charge of this event was Supreme Commander of the Allied Powers General Douglas MacArthur. A few minutes before 9 am the eleven Japanese delegates appointed by Emperor Hirohito, made their arrival in silence.

After General Douglas MacArthur, Admiral Chester Nimitz , and Admiral William F. Halsey, Jr.

greeted the Japanese delegates lead by Foreign Minister Mamoru Shigemitsu, a prayer was said followed by the playing of the "Star Spangled Banner." Then General MacArthur addressed those present and the world with these solemn words;

"We are gathered here, representatives of the major warring powers, to conclude a solemn agreement whereby peace may be restored. The issues, involving divergent ideals and ideologies, have been determined on the battlefields of the world and hence are not for our discussion or debate.
Not is it for us here to meet, representing as we do a majority of the people of the earth, in a spirit of distrust, malice or hatred. But rather it is for us, both victors and vanquished, to rise to a higher dignity which alone befits the scared purpose we are about to serve, committing all of our peoples unreservedly to faithful compliance with the undertakings they are here formally to assume.

It is my earnest hope, and indeed the hope of all mankind, that from this solemn occasion a better world shall emerge out of the blood and carnage of the past, a world founded upon faith and understanding, a world dedicated to the dignity of man and the fulfillment of his most cherished wish for freedom, tolerance and justice."

At 9:04 am Japanese Foreign Minister Mamoru Shigemitsu signed The Instrument of Surrender on behalf of the Emperor of Japan and the Japanese Government, at 9:06 am General Yoshijiro Umezu Chief of the Army General Staff signed the document, then General MacArthur, Commander in the Southwest Pacific and Supreme Commander for the Allied Powers asked British Lieutenant-General Arthur E. Percival and U.S. General Jonathan M. Wainwright to accompany him to the signing table, both men had just been released from Japanese captivity as POWs. General MacArthur accepted the Japanese surrender, "for the United States, Republic of China, United Kingdom, and the Union of Soviet Socialist Republics, and in the interest of the other United Nations at war with Japan."

Upon signing the document General MacArthur used five pens, Lieutenant -General Percival and General Wainwright each received a pen. Then General Chester Nimitz signed the document followed by other Allies.
Memoru Shigemitsu signs the Instrument of Surrender for Japan, officially ending World War II

File:Mamoru Shigemitsu signs the Instrument of Surrender, officially ending the Second World War.jpg. (2022, October 16). Wikimedia Commons, the free media repository. Retrieved 17:04, January 1, 2023 from https://commons.wikimedia.org/w/index.php?title=File:Mamoru_Shigemitsu_signs_the_Instrument_of_Surrender,_officially_ending_the_Second_World_War.jpg&oldid=696736606.

After the ceremony as the sun broke through the morning clouds hundreds of Navy fighter planes and B-29 Superfortress bombers did a fly over of the USS Missouri in wave after wave.

Later after the ceremony, General MacArthur broadcast this strong admonition to the world;

"Today the guns are silent. A great tragedy has ended. A great victory has been won. The skies no longer rain death, the seas bear only commerce, men everywhere walk upright in the sunlight. The entire world lies quietly at peace. The holy mission has been completed. We have had our last chance. If we do not devise some greater and more equitable system, Armageddon will be at our door."

Three years eight months and 26 days after Pearl Harbor, the war was over.

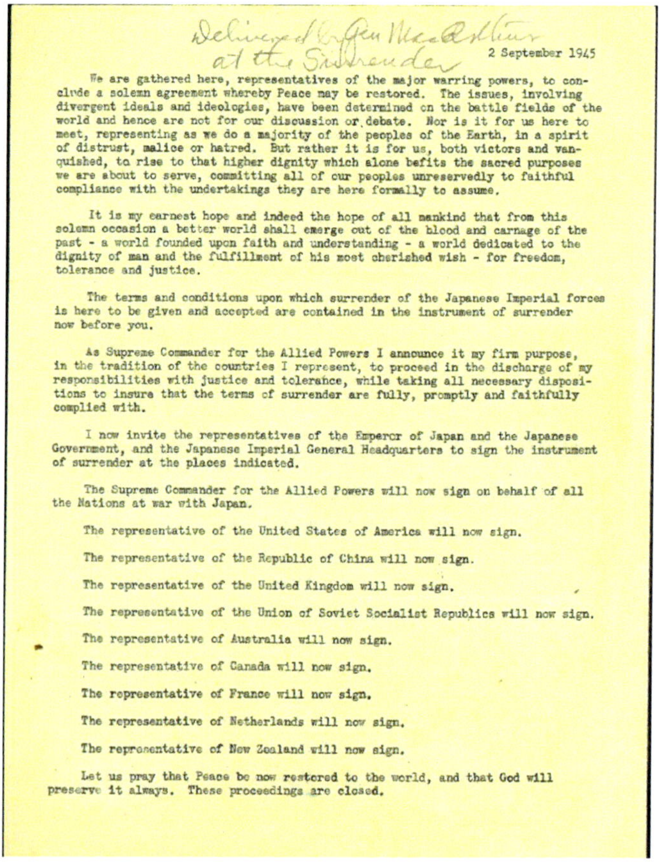
Delivered by Gen MacArthur at the Surrender

2 September 1945

We are gathered here, representatives of the major warring powers, to conclude a solemn agreement whereby Peace may be restored. The issues, involving divergent ideals and ideologies, have been determined on the battle fields of the world and hence are not for our discussion or debate. Nor is it for us here to meet, representing as we do a majority of the peoples of the Earth, in a spirit of distrust, malice or hatred. But rather it is for us, both victors and vanquished, to rise to that higher dignity which alone befits the sacred purposes we are about to serve, committing all of our peoples unreservedly to faithful compliance with the undertakings they are here formally to assume.

It is my earnest hope and indeed the hope of all mankind that from this solemn occasion a better world shall emerge out of the blood and carnage of the past - a world founded upon faith and understanding - a world dedicated to the dignity of man and the fulfillment of his most cherished wish - for freedom, tolerance and justice.

The terms and conditions upon which surrender of the Japanese Imperial forces is here to be given and accepted are contained in the instrument of surrender now before you.

As Supreme Commander for the Allied Powers I announce it my firm purpose, in the tradition of the countries I represent, to proceed in the discharge of my responsibilities with justice and tolerance, while taking all necessary dispositions to insure that the terms of surrender are fully, promptly and faithfully complied with.

I now invite the representatives of the Emperor of Japan and the Japanese Government, and the Japanese Imperial General Headquarters to sign the instrument of surrender at the places indicated.

The Supreme Commander for the Allied Powers will now sign on behalf of all the Nations at war with Japan.

The representative of the United States of America will now sign.

The representative of the Republic of China will now sign.

The representative of the United Kingdom will now sign.

The representative of the Union of Soviet Socialist Republics will now sign.

The representative of Australia will now sign.

The representative of Canada will now sign.

The representative of France will now sign.

The representative of Netherlands will now sign.

The representative of New Zealand will now sign.

Let us pray that Peace be now restored to the world, and that God will preserve it always. These proceedings are closed.

General MacArthur's speech preserved by Naval History and Heritage Command

https://www.history.navy.mil/research/archives/digital-exhibits-highlights/vj-day/surrender/macarthur-speech.html

Commodore Matthew C. Perry's fleet for his second visit to Japan, 1854

File:Commodore Perry's second fleet.jpg. (2020, October 11). Wikimedia Commons, the free media repository. Retrieved 02:21, December 30, 2022 from https://commons.wikimedia.org/w/index.php?title=File:Commodore_Perry%27s_second_fleet.jpg&oldid=4867

Chapter Twenty-Five

The Challenges and Opportunities After Japan's Surrender

After Japanese leaders signed, "The Japanese Instrument of Surrender," on September 2nd, 1945, General Douglas MacArthur, Supreme Commander of the Allied Forces in Japan faced multiple post-war challenges that also presented great opportunities for Democracy.

The Challenges,

Japan's industrial base had been devastated by the constant incendiary bombings of the United States Army's Air Force B-29s for several months. Two of its cities had been completely destroyed by the recent atomic bomb attacks. Multiplied millions of its citizens faced starvation and illness. With these challenges, there was a new political challenge, the potential spread of communism, and the beginning of what the future would call the Cold War. America needed a friend in the Japanese sphere of influence.

The Opportunities,

A chance for Reconciliation with a defeated enemy. A chance to rebuild a devastated Japan through Reconstruction. A time to allow healing through Forgiveness. A chance to change the attitudes of Racial Bigotry between the Japanese and those who had been injured by hate. A time to correct injustices by prosecuting those guilty of War Crimes. A chance to mend the attitude of a defeated Japanese military. An opportunity to Liberate Japan and show the world true Democracy through the positive attitudes of American Occupational Forces toward a conquered enemy.

As part of the surrender, General MacArthur decided not to prosecute Emperor Hirohito for war crimes but allow the Emperor's role to become symbolic. This action went against world opinion, many wanted a more severe punishment, but General MacArthur made his choice which allowed for a smoother transition with the citizens of Japan.

Chapter Twenty-Six

American POWs held by Imperial Japan

POW * MIA Flag: Woodring Wall of Honor and Veterans Park, Inc. Enid, Oklahoma

Liberated POWs raise the American flag at Camp Rokuroshi, Osaka, Japan 1945
Picture: The National Prisoner of War Museum, Andersonville, Georgia

POWs Captives of Imperial Japan

Statistics on the number of American, Allied, and civilian internees held as POWs by Japan can only be estimated due to the destruction of records by the Imperial Japanese and the extremely vast area covered in the Pacific Campaign of War. (One-third of the surface of the world, 64,000,000 square miles). The following information is from the United States Naval History and Heritage Command detailed in their report to Congress dated December 17, 2002, titled U.S. Prisoners of War and Civilian American Citizens Captured and Interned by Japan in World War II. Please remember this is American POWs and Civilians only.

"An estimated 130,000 U.S. troops were captured and became Prisoners of War. Germany held almost 94,000 U.S. POWs, and Japan held over 27,000. Prior to the outbreak of WWII, many American civilians were working and living abroad in areas which were to become combat areas as WWII progressed. According to one report, approximately 4,700 U.S. Citizens were captured and interned
*by Germany, and almost 14,000 U.S. Citizens were captured and interned by Japan."(*18)*

*(*18) https://www.history.navy.mil/research/library/online-reading-room/title-list-alphabetically/u/us-prisoners-war-civilian-american-citizens-captured.html (section,World War II-U.S. Prisoners and Civilian Internees, paragraph 2, lines 9-14)*

A map (front) of Imperial Japanese-run prisoner-of-war camps within the Greater East Asia Co-Prosperity Sphere during World War II from 1941 to 1945.

Attribution: Medical Research Committee of American Ex-Prisoners of War** (https://en.wikipedia.org/wiki/American_Ex-Prisoners_of_War)**, Inc. Research and proof of authenticity by Frances Worthington Lipe.

File:JapanesePowCamps-WWII-front.jpg. (2022, November 20). Wikimedia Commons, the free media repository. Retrieved 15:07, December 23, 2022 from https://commons.wikimedia.org/w/index.php?title=File:JapanesePowCamps-WWII-front.jpg&oldid=708130462.

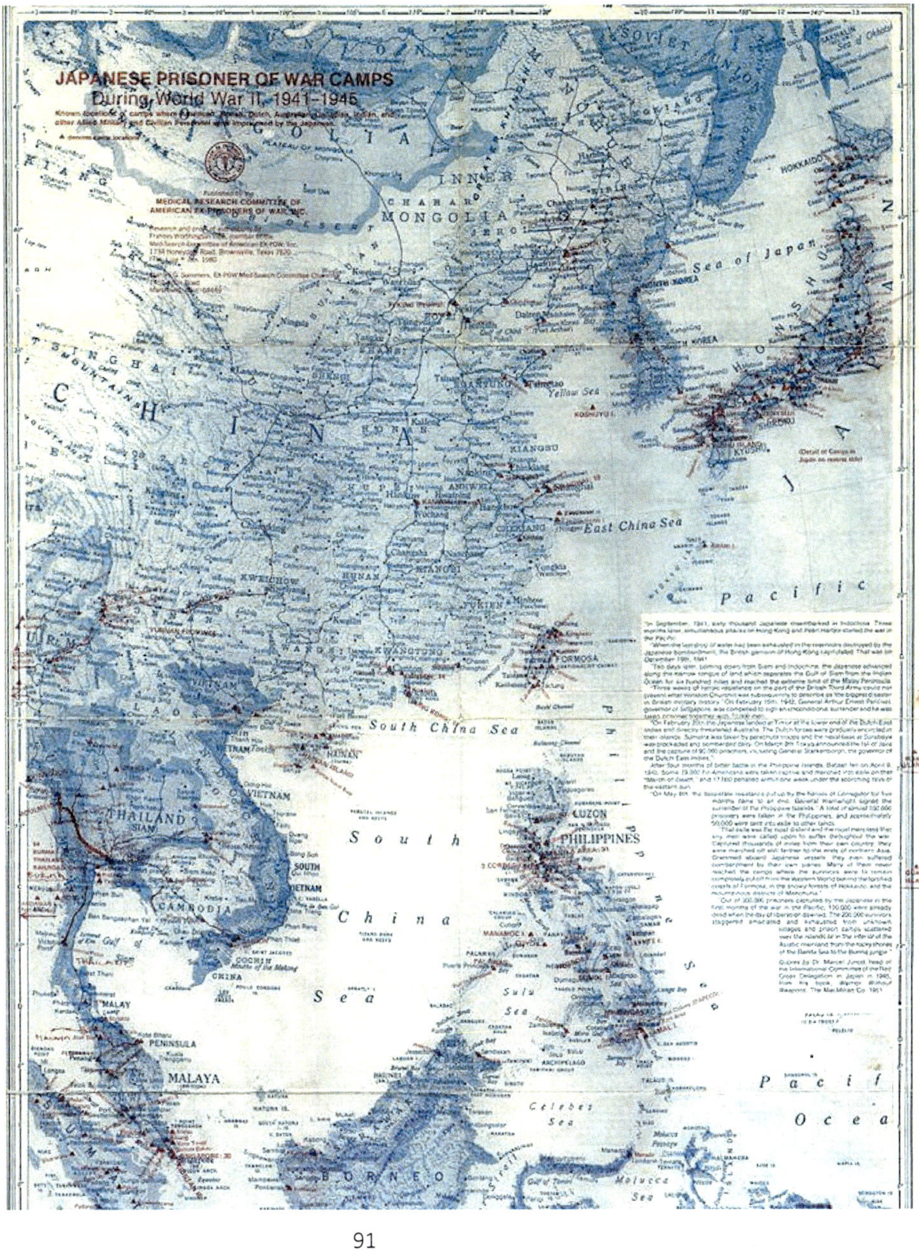

Back of the above map of Imperial Japanese-run prisoner-of-war camps with a list of camps categorized geographically and added detailed map of camps located on the Japanese archipelago.

Attribution: Medical Research Committee of American Ex-Prisoners of War (https://en.wikipedia.org/wiki/American_Ex-Prisoners_of_War), ***Inc. Research and proof of authenticity by Frances Worthington Lipe.***

File:JapanesePowCamps-WWII-back.jpg. (2022, November 20). Wikimedia Commons, the free media repository. Retrieved 15:07, December 23, 2022 from https://commons.wikimedia.org/w/index.php?title=File:JapanesePowCamps-WWII-back.jpg&oldid=708130462.

Chapter Twenty-Seven

The POW Experience-Indescribable

The following short statements and quotes are presented in an attempt to describe the indescribable brutality, torture, slavery, starvation, isolation, and in many cases death that POWs captured by the Japanese experienced before they were Liberated.

NAZI Captivity compared to Japanese Captivity,

"Despite the deservedly brutal reputation of the NAZIs, German captors were more likely to recognize the rights of prisoners assured under the Geneva Conventions. In the Pacific, many prisoners were tortured and many died, either directly or at the hands of their captors, or from starvation or disease.
Captivity under the Japanese was torturous and deadly. POWs were subjected to senseless ,brutality, a direct violation of the Hague Conventions of 1907, a treaty the Japanese had signed."
Statement:The National Infantry Museum, Fort Benning, Georgia

A Quote by an American POW in the Philippines and Japan,

"There were those who dug the graves and those who carried the corpses...There were times when twenty to fifty bodies would lay in the heat, waiting for holes...It was completely mechanical. Once in a while though, you found yourself burying a friend... You just couldn't cover him up without doing something extra." PFC Jack Brady, POW in the Philippines and Japan WWII:
Quote: The National Prisoner of War Museum, Andersonville, Georgia

A Statement about POWs in Camp O'Donnell after The Bataan Death March,

"When Japanese military authorities took the Philippines they assumed that they would be responsible for 40,000 to 50,000 soldiers. Instead, approximately 10,000 Americans and 62,000 Filipinos were forced to surrender. Those who managed to survive the ensuing Death March to Camp O'Donnell discovered that their ordeal was far from over. In the words of one survivor, "The lack of food and water, the rundown condition of the men...all these things would make O'Donnell just one big graveyard."
Statement:The National Prisoner of War Museum, Andersonville, Georgia

A Quote from a POW Chaplin,

"In prison camp it was possible to lose your faith or at least to have it severely shaken. The strange and interesting thing to me was that men under those conditions responded to religion and to God in a very normal way. Prison camp gave them a lot of time to become Christians, not out of fear or because they were suffering, but because they thought it was important to find God.
Captain Robert Taylor, Chaplain
POW in the Philippines and Japan, World War II
Quote: The National Prisoner of War Museum, Andersonville, Georgia

A Statement about families living abroad taken as POWs,

"The lives of thousands of American men women, and children living abroad changed radically in 1942. As war spread throughout the Pacific, U.S. citizens living in the Philippines, China and Indonesia surrendered to Japanese authority. In the Philippines, 3,200 internees were confined at Santo Thomas in Manila, 500 at Camp Holmes, and 2,400 at Los Banos."

Statement: The National Prisoner of War Museum, Andersonville, Georgia

"Hell Ships"

"For the men held in the Philippines during the second World War, Allied victories in the Pacific during the latter part of 1944 were a mixed blessing. To remove the POWs to hard labour camps on mainland Japan, the Japanese government packed them into "hell ships." Tragically, more than 5,000 prisoners died when several of these unmarked vessels were torpedoed by American submarines. It is estimated that more than 5,200 POWs died en route to Japan."

Statement: The National Prisoner of War Museum, Andersonville, Georgia

After the Japanese surrender in September of 1945 American/Allied POWs and Interned Civilians began to be released from the prison camps, for the Survivors of this horrific experience--- life began again.

Chapter Twenty-Eight

Still Searching-Missing in Action

Picture: The Museum of Aviation, Robins Air Force Base, Georgia

Major Edward Cragg's P-38 Lightning "Porky 2" is depicted during an attack on Japanese forces. Cragg shot down a total of 15 Japanese aircraft. His last aerial victory was on December 26, 1943, but he did not return to base and was reported Missing in Action: Painting by Hugh Polder

One of the most heartbreaking experiences any family can endure is to be notified that their loved one is Missing in Action. I remember my Father saying that before battle when soldiers prayed together, they did not pray for themselves but for the soldier next to them. In that same spirit, the Soldier's Creed states, "I will never leave a fallen comrade." This sacred commitment of leaving no one behind on the battlefield is a continuing promise our government makes to our military. Today The Defense POW/MIA Accounting Agency is fulfilling our Nation's promise by actively continuing the search for those missing in action. The following information is from The Defense POW/MIA Accounting Agency detailing their continuing search for those missing from battles in World War II to the present. Below is their mission statement and a map of MIA's detailing the number of missing and the war involved.

Last Update: December 13, 2022

The Defense POW/MIA Accounting Agency's mission is to provide the fullest possible accounting for our missing personnel from past conflicts to their families and the nation. Within this mission, we search for missing personnel from World War II (WWII), the Korean War, the Vietnam War, the Cold War, the Gulf Wars, and other recent conflicts. Our research and operational missions include coordination with hundreds of countries and municipalities around the world.

As this map shows, at present, more than 81,500 Americans remain missing from WWII, the Korean War, the Vietnam War, the Cold War, and the Gulf Wars/other conflicts. Out of the more than 81,500 missing, 75% of the losses are located in the Indo-Pacific, and over 41,000 of the missing are presumed lost at sea (i.e., ship losses, known aircraft water losses, etc.).

Since the renewal of U.S. POW/MIA recovery efforts in the 1970s, the remains of nearly 1,000 Americans killed in World War II have been identified and returned to their families for burial with full military honors. This number is in addition to the roughly 280,000 Americans whose remains were identified in a massive effort immediately after the war.

I implore anyone reading this book to go to The Defense POW/MIA Accounting Agency's website for further information on relatives Missing in Action.

Chapter Twenty-Nine

A Time of Judgement

Ecclesiastes 3:17- I said in mine heart, God shall judge the righteous and the wicked: for there is a time there for every purpose and for every work. (The Bible, NKJV)

Pictures: The National Prisoner of War Museum, Andersonville, Georgia

Picture of "The International Tribunal for The Far East," Imperial Japanese Defendant left corner, General Seishiro Itagaki, right corner, Lieutenant General Akira Muto. Both were sentenced to death by hanging.

General Douglas MacArthur, the Supreme Commander of the Allied Forces, established, "The International Tribunal for the Far East," which convened on April 29, 1946, in Tokyo, Japan. The Tribunal was a military trial set to try the war leaders of The Empire of Japan for their conventional war crimes, crimes against humanity, and crimes against peace, covering the time before and during World War II.

Below is a statement from, The National Prisoner of War Museum in Andersonville, Georgia.

"Following World War II, the eyes of the world focused on the Nuremberg trials and the prosecution of NAZI leaders responsible for the Holocaust. Also tried and sentenced to hang at Nuremberg were German guards who had executed Russian and Allied captives, including those responsible for killing the British prisoners who broke out in the "The Great Escape".

At the same time war crime proceedings in Tokyo tried Japanese officials responsible for the brutal conditions of POW camps in the Philippines and along the Thailand Burma railway. Seven men, known as Class-A War Criminals, were sentenced to death, including Seishiro Itagaki, who was convicted of starving prisoners in the Dutch East Indies, and Akira Muto, Chief of Staff in the Philippines."

Caption: General Seishiro Itagaki, in charge of POWs in the Dutch East Indies, sentenced to hang.

Caption: Lt. General Akira Muto, Chief of Staff in the Philippines, Sentenced to hang.

By the end of 1949 thousands of Japanese war criminals considered to have committed lesser crimes had been tried by tribunals of Allied nations across Asia and the Pacific.

Chapter Thirty

Typhoon Louise

Sergeant Rathel, speaking

After arriving on Tinian Island on the 22nd of June, I was examined by the medical staff and told I would receive better treatment for my injury at the hospital in Guam. A week later I was transferred to the hospital in Guam. By the end of July, my injury was much improved and on August 17th, just after Japan surrendered, I was approved to return to my regiment.

On August 22nd, I watched our Flag fade into the blue horizon of the Pacific Ocean, as I sailed away from Guam for the second time in my life.

While I was recovering from my injury the 77th Infantry Division had completed its mission on Okinawa during the last week of June and during July had moved to Cebu, Philippines; to regroup and resupply. At Cebu the 77th Infantry was preparing for Operation Downfall, the final battle against Japan; fortunately, after the Atomic bomb attacks of August 6th and August 9th, Japan surrendered. By the time I arrived in Cebu, on September 4th, a new mission had been announced, "The Occupation of Japan."

The question everyone was asking upon my arrival at Cebu was, "Where would we be deployed in Japan?" Some rumors mentioned northern Honshu or Hokkaido. To me, both places had cold weather, and I didn't like cold weather.

On September 9th, it was announced the 77th would be deployed to Otaru, Hokkaido, scheduled to arrive on the 5th of October. Hokkaido is the second largest island and most northern prefecture of Japan.

On the 26th of September, the 77th Infantry sailed out of Cebu on seventeen transports en route to the Tsugaru Strait between Honshu and Hokkaido

Map of Hokkaido, Japan, The Tsugaru Strait is located on the lower left of the map above.
File:Hokkaidomap-en.png. (2022, April 5). Wikimedia Commons, the free media repository. Retrieved 21:14, January 7, 2023, from https://commons.wikimedia.org/w/index.php?title=File:Hokkaidomap-en.png&oldid=646916025.

While en route to Hokkaido, the weather began to change and the ocean became very turbulent. On October 2nd, 1945 a tropical cyclone named Louise developed near the Mariana Islands and headed Northwest, intersecting our transport.

Typhoon Louise, October 9th, 1945

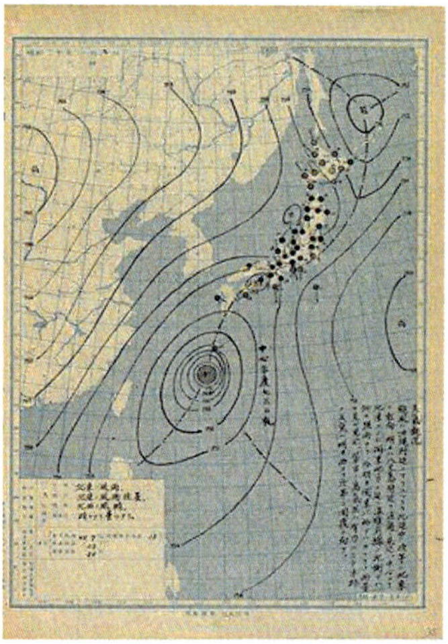

File:Typhoon Louise's Weather map on October 9, 1945.jpg. (2023, January 2). Wikimedia Commons, the free media repository. Retrieved 21:25, January 7, 2023 from https://commons.wikimedia.org/w/index.php?title=File:Typhoon_Louise%27s_Weather_map_on_October_9,_1945.jpg&oldid=722072093.

For three days we tried to dodge the storm as it developed but it was impossible. Eventually the southern part of the storm caught our transports as it moved between Japan and Okinawa. For me
and all the men of the 77th Infantry it was an unforgettable experience, we had to use rope and tie ourselves to the deck of the ship as it was tossed about in violent waves, howling wind and tremendous rain. Many men became seasick. Finally on the afternoon of October 4th, we could see land, Hokkaido, Japan.
On the 5th, of October the 307th RCT disembarked at Otaru and boarded trains for Sapporo, the capital city of Hokkaido, the new home of the 77th Infantry Division.

We later learned that our military based in Okinawa had suffered heavy losses, with many ships being sunk and destroyed. The storm also destroyed the airbase on Okinawa and damaged many planes.

Picture of a damaged ship in Okinawa after Typhoon Louise: The Woodring Wall of Honor and Veterans Park Inc. Enid, Oklahoma

Chapter Thirty-One

The Occupation of Japan

The Victory Flag —This is the first American flag to fly over occupied Japan after World War II, it was placed on top of the radio mast of a U. S. C-47 Skytrain (DC-3) aircraft upon landing at Atsugi Air Base, Japan, August 28, 1945, by Robert Demorest Miller, an air-traffic and air defense controller with the Fifth Air Force: Picture of the actual Flag, The National World War II Museum, New Orleans, Louisiana.

Picture: The National World War II Museum, New Orleans, Louisiana

After the war, Japan was devastated, and the occupation of Japan by the U.S. Military was from September 1945 till 1952. During this time the goal of our Army was to police, maintain order, and help Japan rebuild and heal.

Chapter Thirty-Two

Salute

From the Author,

Near the end of October 1945, my Father along with many others who had served from the beginning of the activation of the 77th Infantry Division in Fort Jackson, South Carolina in 1942, were told they were going home. I'm sure to these men those were beautiful words. They were being replaced by new soldiers who had never been in battle. These new soldiers completed the mission of, "The Occupation of Japan," for the 77th Infantry Division.

The 77th Infantry Division was inactivated on March 15th, 1946.

I find it necessary to Salute the 77th Infantry Division and commend my Father and all the soldiers who sacrificed with him in **The Pacific Theater of World War II.**

According to the 77th Infantry Division Unit History,

In the Four Years the 77th Infantry Division was activated during World War II, they traveled over 25,000 miles, and spent approximately 90 days on ships, with at least 70 of those days in dangerous waters facing possible Japanese submarine attacks. They endured kamikaze attacks, and spent approximately 200 days in battle, killed 43,651 enemy, took 488 Japanese prisoners, and officiated the surrender of 5,543 Japanese in Cebu, Philippines.

A footnote on page 419 of the 77th Infantry Division History, from the Report of Activities, Army Ground Forces, 10 January 1946, lists the 77th Infantry Division as having 3,037 of its soldiers killed and over 10,531 soldiers wounded, with 133 soldiers missing.

The insignia of the 307th Regimental Combat Team,

The motto of The 307th Regimental Combat Team is," Clear The Way"

Much more could be said about these real Heroes, who held, "The Torch of Liberty High," and brought Freedom to the oppressed and captive people of the world.

For my Father his fighting days were not over, to be continued;

Thank you For Reading,

Randy Rathel

About The Author

Randy Rathel lives in Dothan, Alabama is married to Denise Killebrew Rathel. He has two daughters and two bonus daughters. Together they have ten grandchildren. He is retired from the airline industry. They attend Harvest Church in Dothan, Alabama. Randy and his wife Denise love history, genealogy, and traveling to historic places. Randy's first book "Ruby's Memoires," about his Mother's life is loved by many.

***

Made in United States
North Haven, CT
29 April 2023

36006998R20060